DREAM SUPER-EXPRESS

STUDIES OF THE WEATHERHEAD
EAST ASIAN INSTITUTE,
COLUMBIA UNIVERSITY

The Studies of the Weatherhead East Asian Institute of Columbia University were inaugurated in 1962 to bring to a wider public the results of significant new research on modern and contemporary East Asia.

DREAM SUPER-EXPRESS

A Cultural History of the World's First Bullet Train

Jessamyn R. Abel

Stanford University Press
Stanford, California

STANFORD UNIVERSITY PRESS
Stanford, California

Printed in the United States of America on acid-free, archival-quality paper

Library of Congress Cataloging-in-Publication Data
Names: Abel, Jessamyn R., author.
Title: Dream super-express : a cultural history of the world's first bullet train /
    Jessamyn R. Abel.
Other titles: Studies of the Weatherhead East Asian Institute, Columbia University.
Description: Stanford, California : Stanford University Press, 2022. | Series: Studies
    of the Weatherhead East Asian Institute, Columbia University | Includes
    bibliographical references and index.
Identifiers: LCCN 2021019537 (print) | LCCN 2021019538 (ebook) | ISBN 9781503610385
    (cloth) | ISBN 9781503629943 (paperback) | ISBN 9781503629950 (ebook)
Subjects: LCSH: Tōkaidō Shinkansen (High speed train)—History. | High
    speed trains—Social aspects—Japan—History. | Railroads—Social
    aspects—Japan—History. | Technology—Social aspects—Japan—History. |
    Japan—History—1945-1989.
Classification: LCC HE3360.T6 A24 2022 (print) | LCC HE3360.T6 (ebook) |
    DDC 385.0952—dc23
LC record available at https://lccn.loc.gov/2021019537
LC ebook record available at https://lccn.loc.gov/2021019538

Cover design: George Kirkpatrick

Cover image: Tōkaidō-line Bullet Train Opening Commemorative Postage Stamp,
Nihon Kokuyū Tetsudō, 1964.

Typeset by Motto Publishing Services in 10/14.5 Minion Pro

Contents

Acknowledgments

This book, like the bullet train itself, was an idea that developed over many years before construction finally began, and it benefited from the kindness and generosity of numerous people and institutions along the path to its present form. A few short paragraphs are insufficient to allay my debts of gratitude, but they are the first stop on that line.

Throughout years of research and writing, many individuals generously helped with suggestions, critiques, and discussion. Thank you to my writing group partners Kathlene Baldanza, Jennifer Boittin, Anatoly Detwyler, and Maia Ramnath for their incisive comments on work in progress, and to Robert Hegwood, Shuang Shen, and Ran Zwigenberg for generative discussions, reading recommendations, and even sharing materials at various points along the way. I am truly indebted to Leo Coleman for his expert guidance through the growing literature on infrastructure and his partnership in related intellectual endeavors. I also want to thank Sheldon Garon for helpful advice early on as I configured the scope and direction of the project and Tina Chen for her support in pursuing infrastructure as a topic for the Global Asias Summer Institute, as well as a special issue of *Verge: Studies in Global Asias* and several conference panels.

A number of institutions provided venues to present this research. My partially baked ideas benefited from feedback I received from Gerald

Figal and Vanderbilt University's Asian Studies Department; from Shinju Fujihira, Andrew Gordon, Susan Pharr, and the Harvard University Program on U.S.-Japan Relations community; from Paul Kreitman and all of the participants in the "On the Natural History of Destruction" workshop at Columbia University's Weatherhead East Asian Institute; from Anru Lee, Rashmi Sadana, Stéphane Tonnelat, and other participants in the "Ethnographies of Mass Transportation in a Globalized World" workshop at New York University's Center for International Research in the Humanities and Social Sciences; and from Torsten Weber and participants in the DIJ German Institute for Japanese Studies History and Humanities Study Group in Tokyo.

Initial exploratory research at libraries and archives in the United States was supported by a Kent Forster award from the Penn State History Department, a Harvard-Yenching Travel Grant from Harvard University, and a Twentieth Century Japan Research Award from the University of Maryland, College Park. These laid the groundwork for more targeted research trips to Japan, supported by the Japan Foundation and the Northeast Asia Council of the Association for Asian Studies, respectively. I am grateful to both the Humanities Institute and the Center for Humanities and Information (CHI) at Penn State, where residential fellowships provided the time and intellectual scaffolding to work ideas into chapters, and especially to CHI director Eric Hayot for encouraging me to pursue my tentative ideas about the bullet train's connections to information society.

I thank Marcela Maxfield for seeing this project as a fit for Stanford University Press and for her unwavering guidance through the publishing process. I am also grateful to the anonymous readers for the Press: their detailed comments were a model of thoughtful and productive review of scholarship and helped me make this a much better book. Thanks also go to Harrison Cole not only for making an excellent map on a short deadline but also for patience and creative thinking in clearly and elegantly translating my somewhat uncertain ideas for spatial representation onto the page.

As always, my deep gratitude goes to Yamamoto Atsushi and family for fueling my research in Tokyo with not only excellent food, drink, and company but also exciting finds from neighborhood used bookstores. And my affectionate thanks to the Itō-Hirahara clan for their ever-present support, especially Naoko, who always makes sure to remind me that Tokyo and its environs have more to offer than just archives.

Finally, thank you, Jon, not only for the nuts and bolts of helping with barely legible characters and confusing passages, being my first sounding board, and reading everything many times over, but even more for making it possible and worthwhile; and thank you, Ben, for putting up with it all so cheerfully.

Versions or parts of chapters have appeared as "The Power of a Line: How the Bullet Train Transformed Urban Space," *Positions: Asia Critique* 27, no. 3 (2019); "Railway Stations and the Production of Invisible Infrastructures," *City and Society* 32, no. 2 (2020); "Technologies of Cold War Diplomacy: Transforming Postwar Japan," *Technology and Culture* 62, no. 1 (2021); and "Information Society on Track: Communication, Crime, and Japan's First Bullet Train," *Journal of Japanese Studies* 47, no. 2 (2021).

Japanese names are written with the family name first, in accordance with Japanese custom, except when citing works in English, in which case the author's name is written with the name order and spelling used in the publication. For works in Japanese, all translations are mine, except when existing translations are cited.

DREAM SUPER-EXPRESS

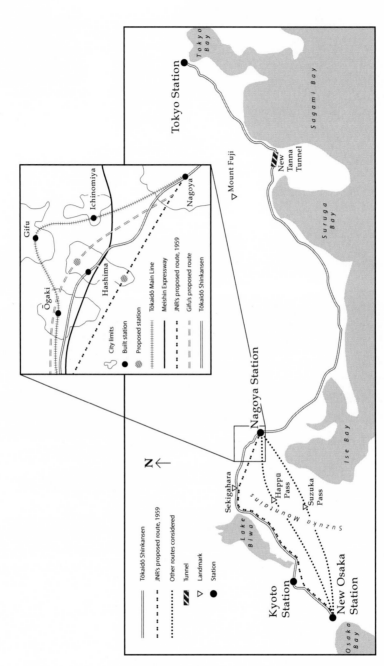

Map of the Tokaidō Shinkansen with alternate proposed routes. Dotted lines for possible routes represent general areas, rather than specific locations. Map by Harrison Cole.

Introduction

Dreams of Infrastructure

AT PRECISELY SIX in the morning on October 1, 1964, two sleek new trains glided simultaneously out of stations in Tokyo and Osaka to capture attention and imaginations across Japan. At every station, the inaugural bullet trains were greeted by flag-waving crowds and children presenting bouquets to the drivers. Bands played rousing marches specially composed to mark this triumphant moment, as local leaders and officials of Japanese National Railways (JNR) gathered to congratulate themselves for their part in building the world's fastest train. People along the tracks stopped what they were doing to wave at the passing train, both witnessing the event and becoming part of the spectacle via the news helicopters that raced alongside it. Even those who were not able to participate in person could join in the celebration from a distance: television and radio audiences experienced the pomp of the opening ceremony and the speed of the world's fastest train vicariously through descriptions and footage of its two-toned blue and ivory cars leaving the station and dashing across the countryside.

But not all observers saw the same thing that morning. The bullet train connected distant cities at unprecedented speed, but it also tore through a region dense with people, industry, agriculture, and history. Opening day was undoubtedly a more somber occasion for those who had been evicted from their homes, had been pressured into selling their land, or were coping with damage to their livelihoods, local environments, and communities. JNR (the public precursor to today's private regional Japan Rail companies) promoted the high-speed railway as a solution to the problem of transportation bottlenecks that were threatening the growth of Japan's industrial economy. But scholars in the emerging field of information studies saw the line in a different light, as part of an emerging post-industrial society. The train was often described as futuristic, but for those who had lived through the Asia-Pacific War, it was also a reminder of the past. Commencing operation just days before the start of the Tokyo Summer Olympics, the new line was not just a means of transportation but one of several new infrastructures that national leaders hoped would showcase Japan's economic and industrial progress and potential to the world. The name of the new railroad was the Tōkaidō Shinkansen, literally the Tōkaidō New Trunk Line, but English-language reports referred to it as the New Tōkaidō Line or by the nickname "bullet train" (*dangan ressha*). That sobriquet was used less often in Japan, but with its evocation of speed and power, it captured the impression Japanese leaders hoped to make on a foreign audience. The domestic press dubbed it the "dream super-express" (*yume no chōtokkyū*), intimating the widespread anticipation and expectations of the Japanese public. This book is about those aspirations and frustrations surrounding this important infrastructure project, the dreams and nightmares of the bullet train.

In recent years, scholars have given careful attention to the political and social impact of the promise held out by infrastructures: the hopes, desires, and visions of a shared future they inspire.[1] Penny Harvey and Hannah Knox demonstrate the power of a road, for instance, to enchant both the workers struggling to build it and the public eager to use it.[2] Brian Larkin draws attention to the aesthetic function of infrastructures, which—alongside their technical function of moving people and things from one place to another—"also operate on the level of fantasy and desire. They encode the dreams of individuals and societies and are the vehicles whereby those fantasies are transmitted and made emotionally real."[3] And although certain types of infrastructure (pipes, wires, roadbeds) are hidden from view, there are some, as John Durham Peters points out, that form "intentional displays of power and modernity," that, like any technological device, have an element of "bling."[4] This function of display is related to what Michael Adas identifies as the modern use of technological expertise as a barometer of national power and the "level" of a civilization.[5] These notions of enchantment, promise, or aesthetic force are all concerned with the power of infrastructures to move people not just physically but also intellectually and emotionally. Taking inspiration from such work, this book views Japan's first bullet train in terms of competing interpretations of the promises it held out to Japanese society in the 1960s; rather than the train itself, it is about the dreams (both good and bad) that it inspired. As an infrastructure that relied on advanced technologies and precision engineering, the bullet train combined the forces inherent in infrastructures and technologies to powerful effect.

These stories of bullet train dreams show that technologies and infrastructures do not have to be entirely new in order to enchant. In fact, though newly built, the line did not represent radical innovation. Its design relied on incremental improvements to existing systems already in use on some JNR lines and methods borrowed from aeronautics and other fields, not a profound technological leap.[6] And it did not create an entirely new infrastructural link so much as enhance an existing one. The original Tōkaidō Main Line connecting Tokyo and Osaka was completed seventy-five years earlier, and with recent improvements, its running time from end to end had been cut from nine hours to under seven hours. With new highways, expanding airline service, and a recently constructed monorail providing fast access to Tokyo's Haneda Airport from the city center, travel between the capital and nearby cities by airplane, bus, or automobile was becoming faster and more convenient. In terms of industrial and technological developments, Japan's airline industry was expanding, and the nation even had the beginnings of a space program. The railroad, in contrast, was seen globally as a sunset industry, a relic of the past in the emerging space age. In this context, it is curious that the opening of a new railroad line, even one that could claim the title of "world's fastest," garnered such rapt attention.

The explanation for this seemingly incongruous national excitement about a high-speed railroad lies not in some universal enthusiasm for three-hour trips between Tokyo and Osaka, but in the power exerted by the very idea of a fast, high-design, high-technology transportation infrastructure. For most people, the line's practical use in their everyday lives was negligible. Even among residents of the cities with stops, many would never ride it.

Tracing the meanings assigned to the bullet train at the historical moment of its conception, construction, and early operation shows that its importance, therefore, came primarily from the ways that it prompted the reimagination of identity on the levels of individual, city, and nation in a changing Japan. The bullet train was built to move people at high speeds from one city to another, but it also moved people's hearts and minds in more subtle ways: it conveyed meanings, instilled feelings, and evoked emotional responses. Those intangible and malleable historical, social, and cultural functions are the subject of this book.

The New Tōkaidō Line became a broadly powerful symbol precisely because its promise (or, for some, its threat) was interpreted in many different ways. Groups and individuals, each with their own distinct but overlapping interests, worked to co-opt, divert, or challenge official narratives about the significance of the line to serve particular purposes. These efforts themselves created new meanings, expanding and diversifying the promises of high-speed rail from JNR's original plan to solve a transportation bottleneck impeding industrial growth to a multifaceted vision that essentially contained something for everyone, even those who suffered from its construction. Harvey and Knox argue that, although painful stories of obstacles, accidents, and frustrations associated with the process of planning and construction are usually written out of celebratory official discourses, in fact, such difficulties and challenges, recounted as "tale[s] of achievement against the odds," themselves contribute to infrastructure's power to enchant by creating modes of individual engagement that are directly related to people's particular interests and concerns.[7] Multiple contrasting narratives together illustrate the various promises embodied or belied by infrastructural projects. Narratives that nudged and questioned the dominant discourse themselves ultimately contributed to the affective power of the bullet train. Exploring these diverse perceptions—the many ways in which the bullet train enthralled people in various walks of life—demonstrates the ways that infrastructure and technology can be exciting without being new and exert power in seemingly unrelated areas of politics, society, and culture.

The bullet train imaginary is similar to the cultural change effected by the introduction, extension, and improvement of railways in the nineteenth and early twentieth century. High-speed rail intensified the feeling of shrinking time and distance that the first railways had created, and it stood, as older railways in both the archipelago and the empire previously had, as a symbol of Japanese power, modernity, and technical achievement. But it differed in ways that were characteristic of its time. In the earlier period, in Japan and its colonies as in Europe and the United States, the development of railroads represented the extension of civilization, the conquest of nature.[8] The bullet train did not push the boundaries of civilization, retracing a well-worn path between the nation's largest urban centers. And in contrast to the nineteenth century, when railways represented Japan's modernization through the import of Western industrial machines and methods, twentieth-century train systems symbolized Japan's own technological prowess. In the period of imperialist expansion, the control and construction of railways in Korea and China were part of the literal extension of Japanese power.[9] The bullet train reflects a different method of regional influence, as railway technology was made part of Japan's postwar rehabilitation through development and technical assistance.[10] Therefore, like previous railway advances, it changed perceptions of space and identity, but in directions that were shaped by its specific historical context.

VIEWING JAPAN FROM THE BULLET TRAIN

Japan in the 1960s was marked by several interconnected vectors of profound change. High-speed economic growth was changing the country's international status, symbolized by the successful hosting of the 1964 Summer Olympic Games in Tokyo, which provided a global showcase for Japan's postwar recovery and prompted extensive infrastructural development (much of which entailed similar aspirations and tensions as the bullet train did). The coincidence of preparation for the Olympics with construction of not only the bullet train but also new expressways together inspired a set of legal changes to facilitate land expropriation. Economic growth was also related to important social changes, such as the demographic challenges of urban overcrowding and rural depopulation and the emergence of the "enterprise society" (kigyō shakai), in which "meeting the needs of the corporation is 'naturally' understood to be social common sense and to be congruent with meeting the needs of all society's inhabitants."[11] In addition, the so-called end of the postwar (proclaimed in a 1956 government economic report to signify a shift in the national economy from basic recovery to a higher level of production and consumption) prompted a turn in debates over the historical significance of Japanese imperialism and, for some, nostalgia for both the hardships and the accomplishments of wartime life.[12] At the intersection of economic growth and industrial development, advances in communications technologies contributed to changes in the structure of the Japanese economy, bringing opportunities for some, but also exacerbating inequalities. The decade also saw a high point of leftist activism in the massive popular demonstrations against the 1960 revision of the U.S.-Japan Security Treaty (Nichi-Bei Anzen Hoshō Jōyaku, abbreviated as Anpo), a protest movement that failed in its primary goals but succeeded on another level by promoting new forms of political action and civic engagement.

The bullet train runs through all of these trends. Just as the New Tōkaidō Line physically cut across the main corridor of Japanese population and industry, connecting the nation's five largest cities, the idea of the bullet train forms a through line in this book, linking five examinations of important political, social, economic, cultural, and diplomatic phenomena of the 1960s. As Langdon Winner argues, technical things like railroads have political qualities; they are tools for building social order. He explains that decision makers indirectly structure and define human associations through two types of choices about infrastructural development: the initial decision to build and the many subsequent determinations, large and small, about specifics such as route and design. The aggregate outcomes of these many decisions both are shaped by and can reshape social relations of power, authority, and privilege. Because of the lasting nature of infrastructure like a railroad line, these changes have a long-term impact on social relations.[13] Dennis Rodgers and Bruce O'Neill demonstrate further that "the material organization and form of a landscape not only reflect but also reinforce social orders, thereby becoming a contributing factor to reoccurring forms of harm."[14] The completion of the bullet train strengthened the very hierarchies of political and economic power that allowed it to blast through contested areas regardless of opposition and to make handsome profits for the rich while further impoverishing the poor.

Polarizing

The bullet train was thus profoundly important to postwar Japanese society as much more than just a symbol of technological achievement and industrial development. As a national project, the new line had a broad impact that extended well beyond the people directly involved. This cultural history of a high-speed railway shows how people used its symbolism for their own purposes, diverting or challenging JNR's claims about it to reinterpret and remake Japanese society and their own place within it. It examines those changing views by tracing the political struggles and cultural contexts surrounding this infrastructure, interrogating the connection of wartime precursors to postwar achievements, and exploring the international significance of what is in practical terms a purely domestic system. Rather than examining infrastructure solely in order to draw out the social and political relationships that form around its operation, this book also uses the public project of the New Tōkaidō Line in order to understand the important events and trends of a particular historical era and to articulate the relationship of mutual influence between infrastructure and culture. While that might be achieved through the examination of any number of important works of infrastructure or material culture, the example of the bullet train provides us with a specific object that inspired great public interest and is therefore surrounded by a rich trove of cultural materials and expansive discourse. It is not the only example, but it is an especially useful one.

Anchoring this history in a widely shared and highly visible object reveals connections among such seemingly disparate topics as diplomacy, popular culture, and evolving attitudes about infrastructure, space, political participation, and economic opportunity. As a major construction project that galvanized the nation (if not in a unified direction), the bullet train shines light on the mobilization of postwar Japanese society and contestation over its future. And as a spark that rekindled a fading international competition to build ever-faster trains, it highlights both Japan's changing international position and the particular ways that a globally used technology was both localized to Japan and then repackaged for the world. In a sense, the bullet train presents central narratives of national history in microcosm: it drew disparate parts of Japan closer together, but at the same time created new divisions and reinforced the marginalization of peripheries in new ways, privileging certain groups while oppressing others, and prompting some to fight for a new place in society. While this is a history of an object, then, it is also a history of a time and place. Focusing on the period of planning, construction, and early operation affords a limited chronology, giving a snapshot of what the train meant at that moment. Of course, the story moves beyond the train itself and extends backward and forward in time from 1964, but it is rooted in that twofold focus of object and moment. The bullet train becomes a window into the 1960s, bringing a novel perspective to a tumultuous moment in Japanese and world history.

Recent studies have addressed various aspects of the bullet train's development. Christopher Hood considers the history of high-speed rail in Japan up to the twenty-first century, providing insight into its symbolism, the role of pork barrel politics in the shape of the network, its financial success and safety record, and its environmental impact. Takashi Nishiyama demonstrates the impact of engineering education on its technical development. Fujii Satoshi examines the relationship between the bullet train and Japanese nationalism. Resonating with some of the topics explored here, Kondō Masataka views a half century of bullet train development from a cultural perspective. And numerous books by former JNR employees present histories of its development from an insider's perspective.[15] This book complements these studies by connecting the story of the bullet train to the historical contexts of the 1960s and by making it not just the subject of the story but also an actor in a broader history.

Dream Super-Express also contributes to wide-ranging interdisciplinary efforts to understand Japanese history and society through material culture. Such a perspective cuts across but also highlights social divisions such as class, gender, and geographic location because the objects of analysis impact all kinds of people, but in different ways. This book foregrounds a shared underlying theme of several recent studies of significant cultural objects and spaces, from cameras, sewing machines, and ramen noodles to Tokyo's commuter rail system and the Ueno Zoo.[16] This approach takes a material object of analysis as a node of connection through which people relate to or interact with machines and nature, other people, and the superstructures of society—culture, state, and other institutions. As a popular object that used advanced technologies and occupied extensive space, the bullet train is a useful junction connecting these various perspectives.

Centering a history of Japan on a technological object requires attention to techno-Orientalist critique, which points to a history of identifying Asian countries, cultures, and people with cutting-edge, even futuristic technologies. Edward Said's explication of the Orientalist view that identified the West with modernity and "the Orient" with the exotic and underdeveloped past was turned on its head in the late twentieth century, as Japan overtook the Western industrial powers both economically and technologically. Looking once more to Asia to find the embodiment of Western anxieties, popular culture increasingly produced images of Japan as the geographical location of the future in the present day.[17] Such representations suggest, to some extent, an admiration of Japanese technologies, but they still exert a dehumanizing effect and define Japan through a Western perspective as an essentialized Other and a threat. This identification of Japan with its technology did not come only from the outside. The bullet train, like other examples of Japanese technological innovation, was marketed to the world as an actual infrastructure, but perhaps more often as an image. In this sense, Japanese diplomats, trade officials, and railway experts helped drive the techno-Orientalist economy, promoting their varied but overlapping interests by selling the idea of Japan as a high-tech nation. The bullet train—as promoted by the railway and other government agencies, as well as private groups and individuals—was drawn into a process of self-essentialization that worked to the benefit of the essentializing subject (who, in this case, is also the object). Those who sought to reinvent Japan as a technological superpower themselves performed a kind of techno-Orientalizing of the self, helping to essentialize the Japanese nation as a producer of advanced technologies and industrial goods.

Many actors participated in that process, some purposefully, others unwittingly. The organizers of the Japanese Pavilion at the 1964–1965 New York World's Fair, who presented both the bullet train and tiny transistor televisions as emblems of a new Japan; the writers of reminiscences that echoed wartime depictions of imperialist Japan as the provider of advanced infrastructures to its puppet state of Manchukuo; the urban planners who envisioned Japan becoming a new kind of "organic society," internally connected and controlled by its communications networks; the local officials who saw in the futuristic train the key to a new metropolitan identity; and the communities who felt it as a threat to their livelihoods: all of these individuals planted seeds that eventually developed into the connections between Japan and technology in the contemporary imagination both at home and abroad. To tell their story is not to fall into the same trap, celebrating Japan as a unique producer of technological wonders, but rather to cast a critical eye on the consequences of that process of identity production and to recognize the history of techno-Orientalism and the role of particular technologies like the bullet train in supporting that image change. In the first postwar decade, the idea of Japanese industry as a leader of technological innovation would have been met with a heavy dose of skepticism from Western audiences. But in the 1960s, the bullet train would join Seiko watches, Sony televisions, and Honda motorbikes to begin changing those attitudes, making Japan into a country that seemed extraordinarily technological. So the history of the bullet train is also a history of the birth of such exoticism.

TIMELINE

Japan's first railroad opened in 1872 between Tokyo's Shimbashi Station and Yokohama, comprising what would eventually become the easternmost end of the original Tōkaidō Line. A section at the Osaka end was operating by 1877. While private companies built lines in other parts of the country, the national rail agency focused on the Tōkaidō, connecting the largest cities. By the time the route was completed in 1889, it was already the core of the emerging industrial economy.[18] Seven decades later, the Tōkaidō region had become heavily industrialized and even more densely populated. Proponents of a new line often cited the same statistics: the Tōkaidō Line accounted for only 3 percent of JNR's system but 25 percent of its passenger and freight traffic; the region comprised 16 percent of the area of Japan, but it contained over 40 percent of the population and around 70 percent of its industry.[19] These numbers showed the line's overwhelming importance to the national economy, which was rapidly gaining strength in the late 1950s, leading demand for transportation to approach the limits of capacity. This trend prompted a decision to build the new standard-gauge, high-speed line alongside the existing trunk line, cutting the running time between Tokyo and Osaka by more than half.

This was not a new idea. One of the significant characteristics of the New Tōkaidō Line was its gauge, the distance between the rails. Though Japanese colonial railways built and maintained standard-gauge railways in Korea and the puppet state Manchukuo, public and private lines in Japan all used a narrow-gauge track, which early planners had deemed the most appropriate to the terrain. Since the late nineteenth century, the Japanese government had intermittently debated whether to move to the wider standard gauge, which would enable safe operation at higher speeds.[20] This recurring debate was finally resolved during the Asia-Pacific War: in 1940, the government approved a plan for a standard-gauge high-speed railroad of one thousand kilometers linking Tokyo to Shimonoseki in nine hours. The railway agency set out to acquire the necessary land and even started work on several tunnels. Construction on the line, dubbed the "bullet train," was abandoned within a few years, as the expanding war consumed all available resources. But two decades later, that modest start on land acquisition and tunnel construction was incorporated into the New Tōkaidō Line. The second (and successful) effort to build a bullet train along the Tōkaidō corridor emerged as a solution to the transportation bottleneck that accompanied expanding economic activity in the 1950s. After completing examinations of the feasibility and effectiveness of several options for increasing capacity to move passengers and freight, the government approved construction of a standard-gauge line between Tokyo and Osaka in December 1958. Construction began just months later, even before the full extent of the line's path was decided, with a ceremony marking the resumption of work on the New Tanna Tunnel at the same site where digging for the wartime line had been abandoned. As that starting point suggests, the efforts made toward building a standard-gauge system in the 1940s were among the many factors that shaped the project in the 1960s.

The architects of the New Tōkaidō Line at JNR aimed for the shortest possible route from Tokyo to Osaka, with the goal of reducing travel time between the terminals, while also considering safety, construction costs, and the convenience of travelers, whose use of the train would, after all, determine its financial success. But railway officials and engineers could not autonomously decide the train's path and station locations. They had to contend with local opposition, as well as nonhuman environmental factors, both of which helped determine both the shape and the many meanings of the line. The intersection of technological capabilities and the physical characteristics of landscapes and cityscapes constrained planners' choices. In some areas, inadequate ground conditions forced adjustments to the route. Safety concerns and efforts to minimize disruption in populated areas impelled the use of elevated tracks. That practice influenced not only the physical characteristics of the line but also popular perceptions of it, as the resulting elevation of the train provided pleasant views, raising its attraction as a site of tourism in itself.[21] The relatively straight line required for high-speed operation could be achieved even through Japan's mountainous landscape thanks to recent technological advances in construction of rail bridges and tunnels.[22] This not only facilitated the train's three-hour travel time but also burnished its reputation as a technological marvel. However, those who enjoyed train window sightseeing experienced the many tunnels as disappointing flashes of darkness. In such ways, environmental elements like seasonal floodwaters or mountains exerted their own force on the line and subsequently were transformed themselves.[23]

JNR also had to contend with local governments, communities, and individuals all along the potential routes. Two localities—Kyoto and Gifu—successfully lobbied for significant changes to JNR's initial plan. Less powerful entities could not alter the path, but many were able to extract smaller concessions from the agency, such as an increase in compensation or agreement to build elevated tracks in a certain area. Even when they were not able to affect the shape of the line, however, protests did color perceptions, as problems and disputes surrounding land acquisition were taken up and reinterpreted from both governmental and grassroots perspectives in popular culture materials.

Once the path of the line was solidified, details of operation were still malleable. For instance, JNR initially planned for the line to carry passengers during the day and freight at night, but ultimately the bullet train operated as a passengers-only service. Service was divided into two categories: the limited express (*tokkyū*), called the "Kodama," and the faster "Hikari" super-express (*chōtokkyū*). The list of station stops for the latter was not finalized until six weeks before opening day, and it was later altered to meet demand. For the first several months of operation, the trains ran at slower speeds to give the roadbed time to stabilize, the Hikari taking four hours and the Kodama five to traverse the full length of the line. But when they eventually achieved full-speed operation, they became the fastest regularly scheduled trains in the world. Not only the speed, but also the smooth ride and sleek design made a strong positive impression on both domestic and international audiences, due at least in part to the promotional efforts of JNR, which included glossy publications with images of the photogenic train in action, test rides for influential people, a color film detailing the line's planning and construction, and cooperation with various media outlets to disseminate images and information.

BULLET TRAIN DREAMS

JNR's promotional efforts yielded the heroic narrative that remains the most familiar version of the bullet train's history, in which the New Tōkaidō Line and Japan's railway industry as a whole represent Japan's leading global position in technological innovation and serve "as a metaphor for a national culture of industry and achievement."[24] The agency praised its new railway system as the fruit of Japan's national spirit and industrial capabilities and the motor of its economic success, emphasizing innovations developed at its own Railway Technical Research Institute and the economic effects of increasing flows of people and goods.[25] Perhaps the most influential of the materials produced by the railway agency was a forty-six-minute film documenting the entire process of construction, from the 1959 groundbreaking ceremony to the inaugural run. The film was shown in Japan (at Shōchiku theaters) and abroad, including a special screening for members of the U.S. Senate in April 1965.[26] It begins with standard images conveying a sense of speed: an oncoming train that suddenly rushes over the camera; a driver's-eye view of the tracks ahead; a focus on the speedometer holding steady at two hundred kilometers per hour; and, finally, a view from afar, encompassing the full length of the train as it races over elevated tracks. The opening credits are laid over the most romantic and exciting views of the train running across the countryside, slicing through rice paddies that extend into the distance on both sides, and disappearing into tunnels.[27]

Having established the thrilling sense of speed and displayed the technical feats of extensive rail bridges and tunnels, the film places the train in the historical context of postwar recovery and the need to expand transportation capacity. It takes the viewer on a high-speed tour of the process of choosing the route and designing what a voice-over notes was "a completely new style of train that could go faster than any other." The narration briefly acknowledges the difficulty of land acquisition but then assures viewers that JNR's representatives "progressed smoothly by gaining the understanding and cooperation of all people concerned." Particular attention is paid to innovations made by JNR, including long rails that made for a smoother ride and special construction methods. Recognizing the importance of aesthetics, the film does not fail to mention "the elegant two-tone light blue and ivory white cars." Several minutes each are devoted to some of the most famous aspects of the train, including the new world record set in a test run on March 30, 1963, and the "electronic brain" that controls most of the driving automatically. The film's tone suddenly changes about halfway through, with dramatic music accompanying a view of the entrance to the New Tanna Tunnel, the site of JNR's most difficult construction challenge. But soon after follows a jubilant scene of tunnel workers shouting "*Banzai!*" for the final blast uniting the two sides of the eight-kilometer tunnel. The film concludes, of course, with opening day and the six a.m. Hikari "ready to leave, carrying the people's expectations and happiness." This new railroad, we are told, was created by "gathering together Japan's brain power and technology and the hard work of many people."[28] Similar ideas and sentiments were presented with great frequency in speeches, interviews, books, and pamphlets, then echoed by newspapers and magazines to the point that Japanese society was saturated with the idea that the bullet train was an epochal achievement of Japanese ingenuity, persistence, and spirit of cooperation.

But the railway agency did not have a monopoly on bullet train symbolism: individuals and groups with diverse interests borrowed or challenged JNR's depiction of a "great national project" that brought prestige to the nation and economic benefits to its people in order to promote their own goals. *Dream Super-Express* considers the significance of the bullet train for people outside JNR and explores the factors that lay behind its symbolic power, illuminating the relationship between infrastructure and identity in early postwar Japan. The new national and local identities that the bullet train helped to create were uncertain and contested. Examining the struggles over its meanings provides a glimpse into the issues and tensions that animated Japanese society in the 1960s and continue to shape contemporary Japanese identity.

A variety of sources provide access to multiple perspectives on the bullet train. Materials produced by the national railway agency yield insight into official aims, and documents from other government institutions, both national and local, reveal points of cooperation and contestation within and between these institutions. But the record extends far beyond government archives. Because there was such popular excitement about the bullet train, many people wanted to write and read about it. Especially in periodicals dedicated to relevant topics, such as travel, railway engineering, or construction, entire issues were devoted to exploring the implications of the new line. But in popular journals and magazines, as well, countless articles, photo essays, and special issues recorded a broad swath of expectations, imaginations, and reactions, as specialists in related fields, famous writers, or anonymous individuals writing letters to a newspaper shared their thoughts. The range of sources for each chapter matches its scope. For instance, Kyoto's city assembly records and narratives about particular communities reveal the impact of planning on a local level, while U.S. publications and institutional sources capture views from outside Japan to address the question of railway diplomacy.

Another entry point into the public imagination is through popular-culture materials, such as television shows, films, and novels. As historian Christina Klein says of postwar American culture, "The exercise of political, economic, and military power always depends upon the mechanisms of 'culture,' in the form of the creative use of language and the deployment of shared stories."[29] Therefore, Klein argues, popular culture is part of domestic and international politics as an inclusive, shared space for constructing and contesting political meanings and social identities. Fictional stories about the bullet train proliferated in the years surrounding the inauguration of this first line and reappeared with subsequent expansion and improvements. These stories may not reveal "what actually happened," but they convey a sense of what people thought and how they understood and grappled with the social changes that inspired the creation of the new line and were spurred by its construction and operation. As Alisa Freedman notes of literature depicting the mass transit systems of early-twentieth-century Tokyo, "authors responded to the contradictions they perceived in Japanese urban modernity, recorded consumer and social patterns often omitted from historical accounts, and exposed the effects of rapid change on the individual."[30] Similarly, Marian Aguiar reads Indian literary representations of trains to discover "the way that the railway as an imaginative object represents the culture, forces, and processes around it."[31] The bullet train system in Japan is, of course, very different from either the earliest mass transit systems of Tokyo or the colonial-era steam trains that dominate views of India, but they all share images of modernity and similarly both shaped and reflected popular attitudes about rail in their own times.

Such materials, therefore, are analyzed here not for their literary merit, but rather as historical documents that provide insight into the times in which they were produced. The bullet train performed a literary role in fictional works in the 1960s and 1970s as a focal point of change and social conflict. The content of the bullet train fiction examined here—a television series, a police procedural in both novel and film versions, a political novel, and an action film—reflect social interests and concerns, but as popular genres they also potentially shaped perceptions of the bullet train and its role in Japanese society differently than either news sources or JNR promotional materials.[32] Tales of the dark sides of bullet train construction worked against the heroic narrative of official discourse, but even such negative portrayals, by representing the human struggles involved, contributed to the bullet train's capacity to enchant. These stories are complicated to use as historical sources, not only because they are fictional but also because of their authorship. Those who created these materials stood outside the disadvantaged groups and communities whose voices and experiences they sought to amplify and represent. But in a situation in which firsthand accounts are lacking, fiction provides a sideways glimpse of what we miss if we never leave the official archives.

The chapters that follow examine several ways in which people reinterpreted, redirected, or challenged official discourses about the bullet train in order to use its symbolic power toward particular goals that often had little to do with the practical matter of getting from one city to another. Chronologically overlapping, they pursue a progressively widening geographical focus, from a single city, to the Tōkaidō region and the entire nation, then the regional space of the former empire, and finally, the global scope of international relations. The first view, a close-up of one city (Kyoto) during the planning stage, shows how two very different groups of people either diverted or questioned JNR's rhetoric about the bullet train to serve their own local interests. Moving forward chronologically, the opening of the new line reveals that a similar dynamic operated on the scale of the Tōkaidō region, changing perceptions of its cities and their relationships to each other. Placing the train in the context of national socioeconomic trends in the years after the new line was built and the high-speed rail system was being expanded highlights its role in debates over the nation's future. Finally, framing the line within international contexts brings out different debate: about Japan's national identity, viewed in terms of its position in Asia and the world.

Chapter 1, "Invisible Infrastructures of Protest in Kyoto," considers the planning process, focusing on two opposing positions in a story of intersecting battles surrounding the route of the bullet train: the little-known campaign by the former capital's local leadership to attract a bullet train station to their city and the more existential struggle of the communities ultimately destroyed to make a path for it. The city and prefectural leadership pressed relentlessly for JNR to change its initial route bypassing Kyoto as part of their ongoing effort to create a new image for the ancient capital by adding a sheen of global modernity to its reputation as a repository of history and tradition. Their success was cold comfort for those whose communities were destroyed, families evicted, or businesses ruined by the construction of the tracks through the city. These groups, too, challenged JNR's bright rhetoric of the bullet train as a great project that would bring national benefits in order to question the structures of democracy in Japan, joining a much broader movement to gain a greater voice for citizens and bolster the power of the individual against state and corporate forces.

Chapter 2, "Reconstructing the Tōkaidō," widens the frame to encompass the full extent of the new line, considering debates among urban planners, general public discourse, and popular culture featuring the bullet train to explain its function in the social construction of space. This perspective highlights the relationship between ideas and infrastructure and sheds light on the dynamics of power over space, including changes not only to physical urban forms, but also in the ways that people understood the cities on the line.

Chapter 3, "Railroad for the Information Society," adopts a national frame, situating the bullet train within fundamental social and economic changes of the 1960s and early 1970s, identified at the time as a transformation from an industrial to an information society. Through debates over urban planning and depictions in popular culture, the new line became a flashpoint that people used to grapple with the problems accompanying those changes and envision a better future.

Chapter 4, "Nostalgia for Imperial Japan," brings the former empire into view with a quick glance backward in time in order to understand the place of nostalgia and memory in the story of the bullet train. This chapter briefly jumps back to the 1930s and 1940s, not to tell a continuous narrative of railway development across the wartime divide but rather to build a foundation for understanding the wartime memories inspired by the bullet train. Two symbolically important wartime trains—the South Manchurian Railway Company's Asia Express and the original "bullet train," the planned (but never realized) express train between Tokyo and Shimonoseki—were the subjects of a 1960s surge of nonfiction reminiscences and fictional stories, which connected the present to a sanitized past, overlooking their dark contexts. Comparing the meanings evoked by high-speed rail in these disparate historical moments shows how the promise of infrastructure was drawn into the construction of public memory of war and empire.

Finally, Chapter 5, "Technology of Cultural Diplomacy," zooms out for the broadest perspective, examining the complicated effect on Japan's international relations of its recognition as a world leader in high-speed rail through examination of the government's use of the bullet train as a diplomatic tool meant to bolster Japan's reputation as a leader in industry and technology. The introduction of the sleek high-speed train—not only the world's fastest but also packed with high-tech features like computerized controls and high-design amenities such as air-conditioned cars and airplane-style seating—helped redefine Japan's position within the global community of nations but simultaneously contributed to growing concern about the threat of Japanese exports on world markets. International displays of Japanese technology were aimed at improving perceptions of Japan and promoting its industrial exports, especially to the United States, while rail-related technological assistance was deployed in rebuilding and strengthening diplomatic relations with developing countries. Foreign observers were impressed by the train, but the government's efforts at cultural diplomacy sometimes inadvertently reinforced old stereotypes.

The New Tōkaidō Line changed over time in response to market pressures, political demands, technological improvements, and economic growth, with stations added, speeds increased, noise reduced, schedules adjusted, and new connections created. Even before it opened, JNR was planning another section of high-speed rail heading westward from Osaka along the San'yō Line to Hakata (in northern Kyūshū), completed in 1975. Subsequent extensions brought the system to the northern end of Honshū and through a tunnel across the Tsugaru Strait, with smaller lines branching out toward the coast of the Sea of Japan. Today the decades-old vision of a nation connected by high-speed rail within a day's travel has become reality, with bullet trains running from Hokkaidō in the north to Kyūshū in the south. These developments did not resolve pressing demographic and social problems, as many planners had hoped. But as the network grew, it continuously reconfigured the space and society of Japan, while changing plans, perceptions, expectations, and representations remade its signification. Additional segments are planned, and a new superconducting maglev line between Tokyo and Nagoya is under construction. Predicted to retake the title of world's fastest, that line is expected to connect Tokyo and Osaka in about an hour's travel time. Though the historical contexts of such efforts are varied, the network as a whole continues to be shaped by the paths blazed for the first bullet train in 1964.

Considering a major infrastructural project like the bullet train from the perspective of the multiple contestations surrounding it helps peel back a presentist understanding to show that such infrastructures, like all of the characteristic elements that made up postwar Japanese society, were not inevitable but rather arose through the process of people assigning values. This view shows why the bullet train means what it does today, why it carries so much power and weight. This book does not seek to cover all aspects of the train but rather simply to consider contemporary attention to the new line from several corners in order to reveal the ways that people used physical infrastructures to build conceptual ones that would help them make sense of their world and reshape it in ways they hoped would make it better.

The train was a polarizing project

There were +'s & —'s on a national & global scale

Advancement of Japan + showing off (publicity)
Symbolic (people used it in their own way

Kyoto wanted the train, and convinced its residents that it would be a positive & then hung them out to dry.

1   Invisible Infrastructures
of Protest in Kyoto

IN HIS SEMIFICTIONAL ACCOUNT of the bullet train's impact on one Kyoto community, Nishiguchi Katsumi depicts a public meeting in which Japanese National Railway (JNR) officials and city leaders explain plans for the New Tōkaidō Line to gathered local residents in a neighborhood near Kyoto Station. In this depiction, the railway representatives present their standard rhetoric about the line in order to convince the audience of its importance, waxing poetic about expected economic benefits and the technological wonder of a train that can travel between Tokyo and Osaka in only three hours. City leaders then add a local twist to JNR's hard sell, focusing on the transformations a bullet train station will bring specifically to Kyoto. The residents are drawn in at first but gradually realize that they are about to lose their homes. Angrily, they begin to raise questions about the democratic character of the decision-making process behind that outcome and the rights of individuals against central government assertions of national interest. They are soon inspired to start pushing back, asserting their right to a role in local decision making.[1] With this roman à clef, Nishiguchi himself contributed to the phenomenon that he was depicting: the varied uses of the bullet train as a symbol and a tool by separate and sometimes competing parties in political battles, both local and national. Just as the people at the meeting challenged the government's rhetoric, Nishiguchi's novel brings to the public eye a narrative that was invisible in the official discourse.

With the specific route of the bullet train yet to be determined, government approval of the project in December 1958 caught the attention of city planners along every potential route between Tokyo and Osaka. The railway agency had its own set of priorities, centered on speed and safety, but within the range defined by those limits, there was some room for input from potentially affected localities. Beginning as an ideal vision of the straightest possible line connecting the terminal cities, the actual route gradually emerged through a contested and negotiated process. This process drew on base possibilities mapped out by a combination of land surveys and technological capabilities, as well as the political tug-of-war between JNR, local governments, community groups, and individuals who lived and worked in the train's intended path. JNR was the most powerful party in such negotiations and disputes, armed with a mandate from the National Diet and the most complete information about technological capabilities and land quality, backed up by expanding legal measures for the expropriation of land for public purposes. Therefore, despite various obstacles along the way, the railway agency ultimately was able to draw a path that satisfied its main priorities. However, the success of a few pressure campaigns demonstrates the potential impact of local interests on central power.[2] And even where they failed to achieve their intended goals, the political contests that emerged all along the line contributed to the ongoing development of popular protest as a mode of democratic action and helped shape the meanings of the bullet train.

Narratives of pressure and protest were largely submerged by JNR's triumphant history. In order to make them visible, it is necessary not only to read both against and along the archival grain but also to find materials with an altogether different grain and layer them all to discover previously imperceptible configurations.[3] This chapter pursues these goals in part by exhuming local perspectives through the regular reports of the Kyoto City Assembly.[4] These handwritten summaries of the discussions that took place among both the full assembly and specific committees—including the JNR Policy Committee (*Kokutetsu taisaku iinkai*)—were created for the purpose of bureaucratic record keeping, not general distribution, and few people at the time (or since) would have been aware of their content. But they provide a sense of the thinking of local officials and the ways in which they hoped to influence decision making by the central government, as well as a glimpse of their responses to resistance among the local population. Officials' predominantly dismissive attitude toward protests led them to give short shrift to that resistance in their record keeping and is thus one culprit in creating another, more difficult challenge: uncovering a history of those who did not themselves record their thoughts and actions. Reading the city assembly reports, as well as a memoir and Nishiguchi's fictionalized account, for traces of protest brings to the surface the invisible infrastructure of various pressures exerted on JNR in its plans for the bullet train.

This situates bullet train planning in the context of two levels at which power was contested. The pressure campaigns launched by city and prefectural administrations to change JNR's initial plans for the line can be seen in terms of the long-standing power struggle between local and central government.[5] Though decision making on the line was solely in the hands of a central government agency, in some cases, city and prefectural officials were able to exert pressure to obtain their desired outcomes. Within Kyoto, the reactions of individual landowners and communities all along the new tracks comprise a part of the trend toward issue-specific local grassroots protests that were on the rise in 1960s Japan. The vast demonstrations against revision of the U.S.-Japan Security Treaty (Anpo) in 1960 served as a model and inspiration for those who, disappointed with the outcomes of bureaucratic decision making or standard democratic processes, sought innovative ways to make their voices heard and promote their local interests against powerful national institutions.[6] Several recent studies have focused on access to and uses of infrastructure as sites for negotiating modes of political participation and citizenship.[7] This chapter views the reverse dynamic: efforts to resist the state's imposition of infrastructure were a crucible for expansion of popular political and civic engagement.

JNR's most prolonged and painful headache involved the planning and construction of the tracks through Kyoto. The process of decision making surrounding Kyoto Station as a stop on the line represents both the greatest success and the more typical failures of local campaigns. Kyoto is a particularly important case to consider in terms of the contest for power between central and local governments, given its long postwar history of leftist leadership.[8] JNR officials sometimes blamed that characteristic for

[handwritten margin note: Goal of this chapter is to uncover true public opinion]

their difficulties in building the line's Kyoto section.[9] However, the city administration did not seem to be approaching the issue from a progressive standpoint. The leaders of Kyoto and the communities fighting to protect their livelihoods against destruction by the bullet train found themselves grappling with a tension that Jeffrey Hanes shows had been at the center of Japanese urban planning since the turn of the century: "city improvement" through infrastructural development versus social reform that treats a city as a community of people, rather than an agglomeration of buildings, roads, and railways, and therefore aims primarily to raise human welfare.[10] But in contrast to the scenario Hanes describes of Osaka's prewar progressive leader Seki Hajime fighting the greed of private landowners in order to plan a more livable city, leaders of 1960s Kyoto were intent on building up the physical city with attention-grabbing monuments of international modernity, for the most part leaving the people to stand up for their own welfare.

Tracing this process helps to recover the contingency of outcomes that, in retrospect, appear to have been inevitable. From our contemporary perspective, it seems obvious that the world's fastest train would stop in Kyoto, the cultural heart of the nation and a major center of tourism. But this outcome was dependent not only on the work of committed political and business leaders and the decision making of the legally and financially empowered national agency in charge (JNR) but also geographical factors beyond any human control. Concerted efforts by Kyoto's city and prefectural leadership garnered the prize of a bullet train station, including a super-express stop. In contrast, popular protests by smaller communities that would be damaged or destroyed by construction had little impact on planning. However, in their efforts to protect their communities, people coalesced into new political formations and developed alternative modes of pursuing their interests both within and outside the existing infrastructures of democratic participation. In this sense, even where protesters failed to change the line's route, they still affected their social and political circumstances, by finding ways to make their voices heard and successfully pressing for fair compensation and assistance with new housing. The strength of the political left in Kyoto's prefectural and city politics figured on both levels of contestation, on occasion complicating the courtship of JNR and, in some small ways, supporting local resistance. Zooming in on this multilayered site of pressures and protests illuminates the factors that determined the outcome of local efforts to shape JNR's decisions about the line, including their broader social and political impact.

The process of determining the bullet train's path through Kyoto involved three groups of people with occasionally overlapping interests but differing and sometimes conflicting priorities and, importantly, very uneven access to the levers of power. JNR drew up plans for the new high-speed railway based on the fundamental goal of creating the fastest possible route while ensuring safe operation and minimizing costs. But the potential of the project to bring benefits or harm to Kyoto as a whole and specific communities within it prompted local administrators and residents to take action to promote and protect their own interests. Administrators—including the city and prefectural assemblies and the chamber of commerce—used established channels to influence central government policies. Communities along the route that was eventually drawn through the city, in contrast, found that the standard infrastructures of democracy were not working effectively for them, and they eventually sought other ways to make their voices heard. These intersecting and conflicting efforts all generated something new. The most tangible outcome was, of course, JNR's new railroad. At a more conceptual level, city administrators used the process as a means of urban identity formation, bringing the bullet train into an ongoing effort to reinvent Kyoto as an "international culture tourism city" (*kokusai bunka kankō toshi*), an identity that combined the city's long-standing association with ancient Japanese history and cultural traditions with its aspirations to a reputation for modern infrastructure and world-class facilities. Local communities contributed to the development of new modes of civic participation. Groups with shared interests built alliances among themselves and with individuals who could provide them with the skills and knowledge to find ways to press for their interests against more powerful political forces.[11]

The aesthetic function of the bullet train line in Kyoto thus varied depending on the ground from which it was viewed.[12] Local leaders wanted to borrow its sheen of modernity and advanced technology to infuse the ancient capital with a new metropolitan identity. In contrast, residents around the station—some forced to move, others now hemmed in on two sides by the old and new trunk lines—saw not a sleek train, but rather the

Officials & civilians had different views
of the train

tracks that tore through their communities. But this view itself had multiple meanings. At first glance, it seems to represent the powerlessness of marginalized communities against both local and national government. However, by emphasizing minor victories in gaining assistance and compensation, those who sought to empower the individual against powerful state and corporate bodies reinterpreted the bullet train as a representation of the ability of the people to stand up for their rights.

COMPETING PRIORITIES

JNR aimed to optimize two sets of tensions. The first was between speed and safety. The goal was to make the trip as fast as possible, but maximizing speed was constrained by the limits of safe operation. The second rested on the bottom line: the principle of minimizing construction costs was balanced by considerations of anticipated passenger usage, which is to say, potential revenue. Major factors affecting this double calculus included geographical features of the land traversed and the extent of local cooperation. As is typically the case, local power holders along the potential routes viewed this national infrastructure in terms of their municipal or prefectural interests.[13] Local governments were sometimes able to turn JNR's priorities to their advantage in arguing for changes to the initial plan. They were most successful in cases where at least one of two factors was present: first, if they were able to situate their demands within the constellation of JNR's interests; and, second, if they were able to mobilize politically powerful institutions and individuals to exert pressure on the railway agency. For instance, in the case of Gifu Prefecture's campaign to shift the route (discussed in Chapter 2), advocates showed that a more northern path than JNR had initially laid out would entail lower construction costs and greater safety. In addition, an influential Diet member stepped in at a decisive moment to advocate for the prefecture. In Kyoto, planners' focus on speed and costs produced a route bypassing Kyoto, which local leaders successfully challenged by making the case that a stop in this important domestic and international tourist destination would contribute significantly to the project's ultimate success in terms of both its technical function and its cultural significance. In addition to lobbying JNR officials with the obvious point that tourists were potential passengers, advocates of a Kyoto stop also convinced individuals in other central government institutions to support their cause through appeals emphasizing their city's status as the heart of the nation and symbol of Japan to a growing audience of international travelers.

These core elements of Kyoto's campaign to attract the New Tōkaidō Line connected the project to an ongoing, multifaceted initiative to rebrand the city. In their pleas that JNR change the route, local leaders promoted a new image of the former capital not only as the repository of Japanese culture, tradition, and history, but also as a global and modern, even futuristic city. And they themselves believed that a bullet train station would be essential to making that image a reality. Their arguments were convincing to at least part of the local citizenry and, more importantly, to decision makers in the central government, who ultimately decided in their favor. These arguments were thus effective in promoting a particular set of urban development plans that included the bullet train. However, as is inevitably the case with infrastructural projects, the policies and plans designed to implement their vision worked to benefit some groups at the expense of others, a circumstance that inspired popular political action. The example of the bullet train's journey to Kyoto Station demonstrates two points. First, urban identities are closely intertwined with transportation infrastructure. And second, those identities are always political. They are consciously constructed for specific goals and used by those in power to promote their own interests and elicit cooperation (or at least compliance) from others. But they also inspire political reactions, engendering new sites and forms of grassroots political activity.

Having promoted their plan as providing a three-hour ride between Tokyo and Osaka, JNR officials were intent on keeping that promise. As a result, preliminary sketches of the route between the intermediate stop at Nagoya and the Osaka terminal all bypassed Kyoto for the sake of a straighter line with no additional super-express stops (see map 1, p. xii). The Suzuka Mountain Range stretches across this path, and JNR initially considered three possible routes across that obstacle. Two of those formed roughly straight lines between Nagoya and Osaka, with Kyoto far out of the way, while a longer potential route passed through Sekigahara, roughly following the path of the existing Tōkaidō Line until after Lake Biwa, where it continued southwest, rather than curving north toward the city. Therefore, it was far from certain that the line would go through the old capital or stop anywhere near it. The possibility that the world's fastest train would bypass

their city spurred local leaders—who were already incorporating the bu
train into their own plans for urban development—into action, beginnir
six-year campaign of pressure on decision makers in Tokyo. The challenging terrain of central Japan helped their cause: the two shorter routes were
eliminated relatively early, when geological tests revealed disqualifying
faults and ground quality, and JNR settled on the northernmost path and
announced a preliminary plan in November 1959.[14] But even that route was
likely to cut well south of Kyoto, where planners were concerned about the
difficulty of obtaining the necessary land through the crowded cityscape.
This meant that any potential stop in that vicinity would be some distance
south of the existing JNR Kyoto Station, itself at the southern end of the
city. To make matters worse, JNR's announcement specified that the super-
express would make only one interim stop, at Nagoya. Only the slower Kodama limited express would stop in the vicinity of Kyoto; the Hikari would
speed past as it ran nonstop from Nagoya to the Osaka terminal.[15]

The local leadership of Kyoto was, to say the least, disappointed by this
announcement. Confident that the train would stop in Kyoto, they had
framed their negotiations with JNR on the basis of that assumption. City
assembly reports show that in March 1958, months before the Diet had even
approved JNR's plan to build the line, Kyoto's leadership was already pes-
tering the agency about the details of construction. The city and prefectural
assemblies and chamber of commerce wrote a joint petition, also signed
by prefectural governor Ninagawa Torazō and mayor Takayama Gizō, and
presented it to the JNR Trunk Line Research Committee, in charge of the
proposal for the new line. The petition begins with a statement of support
for the plan to build a new high-speed railway but quickly turns to the city's
concerns about the specific placement of the station in Kyoto. It takes as a
given the likelihood that the line will come through the city, noting, "This is
our nation's greatest international cultural tourism city and the core of the
Kinki tourism zone, and furthermore, industrial development is ongoing,
with continuing plans for future expansion, so it is naturally being given
careful thought and consideration. However, for the good of future development, we ask that you give special consideration to three points regarding the path through the city."[16] Their demands, which would be repeated
in various permutations for the next several years ad nauseam (and to the
mounting annoyance of JNR officials), were that the new line follow closely
along the path of the Tōkaidō Main Line through the city, that it make use

of JNR Kyoto Station, and that the section within city limits be built on concrete slab elevated tracks. They did not see any need to convince JNR to stop in Kyoto. Rather, they initially skipped that step and jumped to ensuring that the construction and path through their city would benefit the locality and further their own goals for urban planning. But initial meetings with the JNR leadership demonstrated that they were getting ahead of themselves, and the city assembly formally decided in October 1958 to organize an effort to attract the bullet train.[17] A year later, JNR's announcement of a route that would just graze Kyoto's southern periphery inspired the intensification of their campaign.

While questions of speed and construction costs worked against the likelihood of a stop in Kyoto, planners also had to consider the interests and convenience of passengers, whose use would, after all, determine the profitability and success of the line. On this point, tourism was a compelling factor in Kyoto's favor. From very early on in the planning process, at least some JNR leaders felt that Kyoto's status as a tourist destination commanded a stop on the new line. In November 1958, members of a city assembly committee that had originally been formed to promote the elevation and electrification of JNR lines within the city and had taken on the additional task of attracting a bullet train stop went to Tokyo to meet with the railway agency leadership and the minister of transportation. They presented a written request and supporting materials to JNR vice president Ogura Toshio, who responded by noting various obstacles and uncertainties. However, he conceded that a stop would contribute to demand, noting that "Tokyo and Osaka have large populations, but for tourism, Kyoto is number one."[18] Ogura again showed his sympathy for Kyoto's cause in two July 1959 meetings with city assembly representatives, saying, "If you consider the Shinkansen as something for all of Japan, I think it is impossible to exclude Kyoto"; he further noted the need for JNR to focus on building for tourism, especially in anticipation of an expected boom connected to the 1964 Tokyo Olympics.[19] He later assured city leaders that "there cannot be a Tōkaidō Line without Kyoto."[20]

But other JNR leaders were more skeptical. On the same late-July visit, city assembly representatives met with another important figure in decision making. Ōishi Shigenari was the agency's executive director in charge of the project and, with JNR president Sogō Shinji and chief engineer Shima Hideo, one of the so-called Shinkansen Trio (*Shinkansen sanbagarasu*), who were given that nickname for their leading roles in developing the plan for the new line. In that meeting, Ōishi questioned the usefulness of a Kyoto stop in the eyes of high-speed rail planners, for whom Osaka and Kyoto were essentially one single entity. He added that including another stop would increase the time of the journey, "and the Shinkansen will lose its meaning."[21] The issue of time and costs pushed decision makers toward a Kodama-only station far south of the city center, but Kyoto's role as a tourist draw suggested that a stronger connection to the city would add to the line's success. Kyoto's leaders focused, therefore, on the latter point in their efforts to bring the line through the city.

Over the course of about six years of planning and construction (from 1958 right up to the weeks before the line's inaugural run), the city assembly cooperated with the prefectural assembly and the chamber of commerce to make repeated (and repetitive) written and oral petitions to JNR officials, leaders of the transportation and construction ministries, and Diet members, including both Kyoto's representatives and the leadership of the transportation committee.[22] Local leadership was not monolithic, but these institutions joined forces to push for their most pressing shared goals: (1) bringing the bullet train through Kyoto, including a stop on the super-express, (2) locating the bullet train station alongside the existing JNR station, and (3) ensuring not only that the bullet train tracks would be elevated but also that JNR would elevate the existing Tōkaidō Line within city limits with a simultaneous construction project. With the exception of simultaneous elevation of the existing line, their efforts were ultimately successful. But their stubborn pressure for these goals, alongside the difficulties JNR had in negotiations with communities around the station, nearly cost the city the coveted prize of the super-express stop. Thus, this is a story of the contingency of historical decisions about infrastructure that transform not only the landscape but our views of the world such that today it seems almost unthinkable that planners would have considered building a route that did not run through Kyoto, important as it is in Japanese history, culture, and contemporary tourism.

DRAWING THE LINE TO A NEW ANCIENT CAPITAL

Kyoto in the early postwar years was a city in search of a future. In her analysis of Kawabata Yasunari's 1961 *Old Capital* (*Koto*), Alice Tseng describes a city struggling to rebuild and re-create itself. It does not thrive, but rather "perseveres, with grace. . . . Kyoto has become a shell of its monumental pasts. . . ."[23] Having survived the 1868 loss of the imperial residence by remaking itself for the new emperor-centered society as a site of imperial commemoration and history, the city was again stripped of its identity by Japan's military defeat and subsequent dismantling of that society. Its survival would depend on another reinvention in harmony with the new values of Japan's postwar society.[24]

Anxiety about Kyoto's future made local officials fairly desperate not only to lure the bullet train but also to ensure that both the regular Kodama and the super-express Hikari would stop there. The stakes were more than just faster access to Nagoya and Tokyo. A key element of urban planning was the promotion of tourism, the city's major industry, and local leaders certainly anticipated that a bullet train stop would give a boost to the local economy, but that was not their only concern. One of the most important factors was the potential impact of a super-express stop on Kyoto's status and reputation. Examining the impact of high-speed rail on Japan's cities during the initial years of operation, urban sociologist Isomura Eiichi found it had created an urban hierarchy in which cities with a bullet train station gained cultural, economic, and administrative power over those without one, and "Hikari-level" cities—those with a super-express stop—gained importance over "Kodama-level" cities, which were bypassed by the fastest trains.[25] These conclusions would have been no surprise to localities whose fortunes had already risen or fallen with previous changes to the railway network, and Kyoto's leaders, like many others, foresaw that the bullet train would have even greater meaning for the cities it connected (or neglected).

This dynamic gave the bullet train an outsized role in determining Kyoto's future, at least in the minds of its leaders. During the city assembly's October 1958 discussion of a campaign to attract a station, one member of the JNR Policy Committee warned, "If the Shinkansen does not come through here, Kyoto will inevitably fall to the level of a second-rate city. On the other hand, if it stops at the existing Kyoto Station, it will be an opportunity to realize elevated tracks, so we must put all our strength into attracting it."[26] The sense among the leaders of this campaign to attract the line was that JNR's decision would either condemn their city to second-tier status or, conversely, provide the impetus for a long-desired

[margin handwritten note: The train validated a city's status]

piece of urban improvement: the construction of elevated tracks to replace the surface tracks of the Tōkaidō Line within the city limits, an infrastructural improvement that would increase both safety and mobility within the city. Thus, securing a bullet train station became an essential part of the assembly's primary goal for 1959, which they defined as urban development aimed at making Kyoto a "global culture city."[27] This goal also entailed attracting a new international conference center, along with other related improvements to the cityscape and infrastructure.

One artifact of this campaign was the rhetoric that Kyoto was the world's window on Japan: the nation's representative international culture tourism city, combining old and new with a modern station and the latest railroad technology in the home of the ancient capital. City leaders argued that the bullet train should come to Kyoto because of these qualities, but they also believed that having the bullet train stop there would itself promote these qualities, resulting in a circular argument. The aspirations and anxieties of city leaders were wrapped up with their campaign to attract the bullet train. Exuding an aura of ultramodernity that planners hoped would overflow its tracks to color everything it touched, the bullet train would help transform Kyoto into a modern global city.

This new image for the city was not invented for the bullet train campaign. Official pursuit of the goal to make Kyoto into an international culture tourism city was legally codified in 1950 through one of several "special city construction" laws (*Tokubetsu toshi kensetsu hō*) passed around the same time in order to promote tourism as a national economic driver by highlighting the particular attractions of certain cities and stipulating the maintenance and development of related facilities. The law delegated implementation to the mayor of Kyoto, thus encoding construction of an international culture tourism city as an explicit responsibility of local government.[28] The city leadership followed up on this law with the adoption of a new city charter in 1956, which made the concept of the international culture tourism city the very foundation of civic life. The new charter's preamble stated: "We citizens of Kyoto, with pride as citizens of an international culture tourism city, in order to beautifully enrich our Kyoto, put forth this charter as the standard to which citizens are obliged."[29] The five-point charter then called on the populace to create a beautiful city; to maintain a clean environment; to display good customs and manners; to protect the city's cultural assets; and to extend a warm welcome to visitors. Tourism, of course, was not new to Kyoto's identity; the city's cultural and historical assets had been the biggest draw for visitors since the emperor moved to Tokyo in 1868. The important new element for postwar Kyoto was the "international" label, which signaled two of the metropolitan leadership's urban planning goals: further increasing the city's already growing popularity with visitors to Japan from abroad and ensuring that it lived up to global standards for a modern city. The idea behind these plans was that explicitly adding an international aspect to the city's identity would represent the direction of postwar Kyoto.[30]

In this sense, the title of international culture tourism city was more aspirational than actual. In fact, the repeated statement of these goals reflects local leaders' anxieties about aspects of their city they felt were less attractive, even ugly or embarrassing, and plans to improve the city combined construction with projects to update or "clean up" such areas. Looking back at their accomplishments of 1959, the city assembly celebrated their resolution to direct new urban development toward building a global cultural city. They dubbed 1960 "the first year of 'a golden age'" and anticipated that it would be the "implementation year" for such plans, especially infrastructure building and other urban improvements.[31] Alongside the campaign to attract the bullet train, city assembly leaders were busy petitioning the central government to choose Kyoto as the site for the construction of a new state-sponsored international conference hall, which they hoped would bring major conferences, concerts, and other activities characteristic of a modern, global city. They also planned the cleanup of "slum" (*suramu*) areas by tearing down existing structures and building modern housing, including the "cleaning" (*jōka*) of the *burakumin* district of Sūjin, just east of JNR Kyoto Station, in order to "beautify the front entrance" to the city.[32] The 1962 reelection of Mayor Takayama was interpreted as a referendum on his policies and plans "to build a modern city harmoniously fused with a thousand-year-old history and tradition," including not only "unsparing cooperation" on construction of the bullet train, the Meishin Expressway, and the international conference center, but also more mundane work, like road and sewer maintenance, waterworks expansion, and other concrete activities.[33]

This rebranding of Kyoto became the crux of urban planning policy and the center of the bullet train station campaign. Various permutations of

the phrase "international culture tourism city" appear everywhere in p[ub]-
lic discussions of the bullet train route, as well as in pleas and petiti[ons]
to JNR. But in order to appeal to a national agency working on a national
project, it was important to connect local concerns with national-level matters. This resulted in the fusion of the rhetoric of the international culture
tourism city with the notion of Kyoto as the heart of Japan's traditional culture and, therefore, representative of Japan itself. The significance of the
"international" element of Kyoto's campaign is clear in JNR vice president
Ogura's comment to visiting city assembly representatives that "the new
line will become an advertisement for Japan."[34] The frequent mention of
foreign visitors suggests that the international gaze was important to this
argument. The fact that a Kyoto stop would attract passengers helped convince JNR decision makers that it would be a sound investment that would
add to the success of the new line. And emphasis on Kyoto's mixed character as not only the ancient capital but now also an international city suggested that it would contribute to external re-evaluations of Japan.

Ultimately, these arguments hit their mark, and in January 1960, JNR
president Sogō announced during a visit to Kyoto that the bullet train
would most likely stop at the existing Kyoto Station and run through the
city on elevated tracks. Visiting with city and prefectural leaders, he lis-
tened to local demands and, in turn, asked for cooperation in helping JNR
to complete land purchases quickly, in order to have the line built and running by the time of the Tokyo Olympics. In a press conference, he extended
that request to the general public, asking for their understanding of the
project's national importance. He also stated at that time that the super-
express would not be stopping in Kyoto but left open the possibility of rethinking that decision.[35] City leaders and JNR both used the same rhetoric
of national importance to promote their interests in local-national negotiations. And even at this moment of victory for Kyoto's battle to attract the
bullet train, JNR dangled the possibility of the super-express as an incen-
tive for local cooperation.

Indeed, this became a new focus for Kyoto's negotiations with JNR. Having succeeded in drawing the line to Kyoto Station, local officials returned
to their initial approach of using bullet train planning as an opportunity to
achieve other rail-related goals, especially making Kyoto a super-express
stop and ensuring elevation of the existing Tōkaidō Line tracks through
the city.[36] In addition, they began pushing JNR to bear a large share of the

necessary construction costs for the station area. As early as April 1960, just a few months after the decision in favor of Kyoto Station, a joint meeting of representatives of the chamber of commerce and the city and prefectural assemblies agreed to demand that JNR bear all costs for road construction related to the new line, contrary to an existing agreement with JNR and the Construction Ministry.[37] They ultimately failed to convince JNR to elevate the existing line alongside bullet train construction. And railway officials continued to put off the decision on whether the Hikari super-express would stop in Kyoto, a question that moved to the center of Kyoto's pressure campaign for the next few years. But the city's persistence in pushing JNR to pay for urban improvements related to the bullet train contributed to tensions with the rail agency, which announced in late 1963 that the fastest Hikari trains would stop only in Nagoya en route between the terminals; Kyoto would be a "Kodama-level" city. The city's lobbying machine had slowed significantly, but in reaction to this news, it roared back into action. The city assembly's JNR policy committee members resumed their schedule of frequent trips to Tokyo to make their case to the rail agency leadership and other central government officials. It eventually became clear that the decision to bypass Kyoto resulted in part from JNR officials' frustration with Kyoto's pattern of uncooperative behavior.[38]

As city officials pressed JNR for a firm answer about the super-express stop, the rail agency waffled and continually put them off, saying a decision had not yet been made. Even into the summer of 1964, city officials were still fretting over the question, visiting Tokyo in July to plead their case one last time. Meeting with several officials over the course of two days, they received the same response from all: before they can consider a super-express stop at Kyoto, they must first resolve disagreements with the city and its residents regarding responsibility to provide housing for evicted residents and for construction in the station area and along the tracks.[39] As late as August 4, just eight weeks before the train's inaugural run, the chair of the city assembly's JNR Policy Committee, reporting on his recent visit to the railway headquarters, gave the prospects for a super-express stop fifty-fifty odds.[40] The decision finally came down on August 18: the Hikari would stop in Kyoto, after all. The impasse was resolved in eleventh-hour negotiations between Mayor Takayama and Murase Kiyoshi, the head of JNR's Osaka Trunk Line Construction Department. Ultimately, a promise from city officials to cooperate in resolving the "outstanding issues" led to a reversal by JNR, and Kyoto station was readied to receive the inaugural bullet trains on October 1. In other words, the city would stop pressing JNR to cover costs for construction related to the bullet train and the relocation of people evicted for it.[41]

The decision prompted widespread relief and celebration among the city's leadership, as well as public ruminations on why the decision had been so difficult. The local newspaper *Kyoto shinbun* provided a detailed account of the decision-making process, explaining that sources within JNR emphasized considerations of running time and blamed their reluctance on concern about breaking the promise of the catchphrase "three-hour express." Given their initial plan of travelling 515 kilometers at an average of 170 kilometers per hour with one interim stop at Nagoya, they insisted, there was no leeway for an additional stop at Kyoto, which they feared would bring the journey to as much as five minutes over the three-hour goal. Some voices within the railway agency also blamed Kyoto's divided government (the prefecture had a Socialist governor in Ninagawa, while the city had a mayor elected with support of the conservative Liberal Democratic Party [LDP]) for complicating local cooperation.[42] This theory may be based on a long suspension of progress after the governor stepped in to halt construction on a new rail bridge over the Kamo River on March 5, 1961. Construction had just begun the previous month and was held up for nearly a year while JNR negotiated with local groups over compensation. Governor Ninagawa approved the restart on October 25, 1961, and construction resumed the following January. Though the prefecture's explanation for the interruption of work on the rail bridge was that they questioned some of the technological aspects of its construction, it seems also to be related to pressure from the prefectural assembly not only to ensure that the Kyoto section of the bullet train would be on elevated tracks, but also to extract a promise from JNR to double-track the local San'in Line.[43] In fact, throughout the 1960s, Ninagawa's administration pursued policies that would favor traditional, small-scale industries and resisted central policies benefiting large industry and high-speed economic growth.[44]

JNR denied the widely held suspicion that they had always intended to have the super-express stop in Kyoto but were holding the decision hostage in order to pressure city leaders. However, Kyoto's persistent pressure, combined with its reluctance to facilitate land purchases and efforts to shirk costs, caused the city and the rail agency to develop "an emotional opposition."[45] This began very early in the planning process. One JNR official, reacting to visiting assembly members' persistent petitioning, told them, "Kyoto has a bad reputation. People say that you have too many demands, and as an arrogant tourist city, you will push us around."[46] He pointed out that by giving in to Kyoto's demands in the first place, JNR had created a more difficult job for themselves and suggested that city officials return the favor with greater cooperation.[47] That such cooperation was not forthcoming hardened feelings against Kyoto within the rail agency. Indeed, transportation minister Matsuura Shūtarō stated publicly that the city's uncooperative attitude regarding land purchases and construction had brought about the debate on whether there should be a super-express stop there. But, suggesting the effectiveness of the city's international cultural tourism argument, he continued, "Kyoto is a historic tourist city, and there will be many foreigners riding the super-express, so from the point of view of international tourism, we must not get caught up in emotions. I will make an effort to have it stop here."[48]

Kyoto's rhetoric of the international cultural tourism city was not only convincing to JNR. It also took hold in the popular imagination, changing the way the people of Kyoto themselves understood and envisioned the bullet train's significance for their city. Analyzing JNR's decision in a public forum, scholars echoed the city assembly's two main arguments for a Kyoto stop in assessing its potential impact: that it would raise international perceptions of Japan as a nation and help maximize ridership. Ritsumeikan University professor Maeshiba Kakuzō stated, "A Kyoto station stop is natural. The group from the IMF General Assembly and the World Bank all came to Kyoto on a test run of the super-express. How will they feel if they come to Japan again and it doesn't stop in Kyoto? The stop is less for the sake of Kyoto than for Japan, of course." Tourism Women's University professor Tonomura Kanji responded with an emphasis on the importance of attracting passengers: "Apparently JNR wanted to fly past Kyoto so they could run in three hours, but the purpose of the super-express is not just running, but rather carrying people."[49] His implication was clear: unless the line was useful to passengers, speed alone was meaningless.

Anxiety about Kyoto's status helped drive the campaign for a bullet train stop, and attaining that goal brought a sense that having one of the four Hikari stations cemented Kyoto's position as one of Japan's leading cities. City and prefectural leaders brought the rhetoric developed for JNR to the public, telling newspaper reporters that the last-minute decision to include Kyoto on the super-express schedule indicated national recognition of Kyoto's importance. They predicted that the super-express stop would provide "a close-up of Kyoto's international value," and that, as a result of JNR's decision, Kyoto's "grooming as an international city is complete."[50] Mayor Takayama echoed these sentiments on the eve of opening day with a statement included in an advertisement for Kyoto businesses, placing the bullet train stop in the context of a broader set of policies to build a new international Kyoto. He wrote that while they were celebrating the birth of the bullet train, the city's new International Hall was nearing completion, and with these accomplishments "Kyoto, which has beautiful scenery and important history . . . is in the midst of a leap forward as an international culture tourism city in both name and reality."[51] The phrase "in both name and reality" hints at the anxieties that remained about this new Kyoto.

In some ways the bullet train actually threatened to undermine the city's new image, not only by bringing more people to the city and juxtaposing the relatively shabby station area with the sleek mode of transportation by which they had just arrived but also because of the new proximity to Tokyo—now accessible even as a day trip—which would encourage comparison of the two cities. In the same advertisement containing the mayor's praise, the chairman of a local business organization warned that "proximity to global city Tokyo will probably negate Kyoto's distinctiveness. Therefore, Kyoto must make all plans from a national and global standpoint, while conversely, regarding tourist attractions, we must maintain Kyoto's distinctiveness. This will lead to a literally global Kyoto."[52] And in a roundtable discussion published in the local daily *Kyoto shinbun*, the newspaper's editor, Tsuji Shūji, noted the low percentage of paved roads in Kyoto relative to other world cities and asked, therefore, "Can one call it things like a cultural tourism city or 'Global Kyoto'?"[53] As these two comments suggest, Kyoto boosters feared both that the bullet train would accelerate cultural homogenization and that Kyoto might not meet international standards of a modern city. For instance, passengers entering the city from the east would pass through the type of "slum" neighborhood that the city had hoped to "clean up," creating a jarring contrast with the clean modernity of the train.[54] These anxieties led to calls for improvements to the station facilities, its surroundings, and the city as a whole to reinforce its emerging identity as modern and international, while maintaining its particularity as the ancient capital.

One focus of such work was the station itself, which had been carefully designed to showcase both "international" and "traditional Japanese" elements. The local newspaper evaluated the new station building in precisely these terms. As opening day approached, several articles celebrated the modernization of the station and its environs. One of these, part of a series focused on the cityscape of Kyoto, was published alongside the news that the super-express would indeed stop in the city and, therefore, doubled as a celebration of that decision. It traced the recent history of modernization of the area surrounding the station, beginning with the construction in 1961 of "the only automated post office in Asia."[55] While the new central post office was a beacon of modern architecture and technology, the article noted that the surrounding neighborhood was still characterized by old tile-roofed houses among buildings stained with soot from before the electrification of the original Tōkaidō Line, eliciting memories of steam trains belching smoke as they chugged their way through the city. But now, the super-express bursting through would bring with it "a wave of modernization" that would make the Kyoto station front "literally the front entrance of the global tourism city 'KYOTO.'"[56] Local figures voiced similar ideas. In a published conversation between JNR Kyoto Station Master Takai Hitoshi and Shishido Keiichi, the president of the Railroad Friends Society Kyoto Branch (and an engineering professor at Kyoto University), Takai agreed with Shishido's comment that the new station was pretty nice, saying, "Sure, because it is the front entrance to an international tourism city. It cost about 2.4 billion yen to build, and some say it is nice enough to have an evening party there. Marble slabs imported from Italy for 15,000 yen each were used for the columns, but what really worried us was how to bring alive the rustic simplicity of the ancient capital."[57] The modern architecture of the station was reflected in its massive 39,000-square-meter, three-story reinforced concrete structure, while the latter effort involved planning a rock garden and a gold and brown interior color scheme that was "appropriate to the ancient capital."[58] To maintain Kyoto's dual character, the station, too, had to exude both a modern, international style and traditional Japanese elegance.

For city planners, the bullet train's aura as the pinnacle of global railway design would contribute to their efforts to reshape Kyoto's metropolitan identity. But the "international culture tourism city" was an urban identity promoted by the local leadership. That image of Kyoto was not equally important to all citizens. Of course, building new infrastructure through a densely populated city necessarily requires some people to sacrifice more than others. These elements of Kyoto's identity were least relevant to those for whom the personal costs would be highest: people evicted from their homes and workplaces or condemned to life under the shadow of busy elevated tracks. Celebrations of the assembly's success in bringing the line to Kyoto and securing a super-express stop obscure the protests, hardships, and loss among segments of the population who would experience more harm than benefits from the bullet train coming through Kyoto. Envisioning the future, they saw not a streamlined train evoking speed but rather the dust of construction, the oppressive shadows and obstacles formed by the infrastructure of the overhead tracks, and the vibrations and noise of constantly passing trains. For them, therefore, the significance of the bullet train was primarily negative, as an emblem of their continued powerlessness, even within Japan's new democratic system. As Penny Harvey and Hannah Knox found in the case of road building in Peru, for communities that felt betrayed by their democratically elected representatives, infrastructure's promise of integration and inclusion came up against a very different narrative "of disillusionment and disempowerment in the face of bureaucratic control, fraud and corruption."[59] In Kyoto, that experience inspired new modes of political participation and the creation of alternative democratic infrastructures.

INFRASTRUCTURES OF PROTEST

The local communities that would be most adversely affected by bullet train construction included, perhaps not surprisingly, some of the most economically and politically marginalized people of Kyoto. The land to be appropriated for the tracks and expansion of the station was located in mostly poor districts, some with inhabitants who had long been victims of discrimination and institutionalized inequities, including an area that was formerly an enclave of the historically outcast *burakumin* and neighborhoods with a heavy ethnic Korean population. The residents of these neighborhoods were the least likely to use the bullet train once it was built: few, if any, had the leisure time, business reasons, or economic resources for high-speed trips to Tokyo. But they were being called upon to sacrifice the most for its construction. As much as they may have been excited about the concept of the world's fastest train and willing to do their part to support a major national project, the unfair distribution of its costs and benefits was hard to accept and inspired new forms of political participation and activism.

In this sense, the bullet train was drawn into the national struggle over Japanese democracy. Examining the contemporaneous but vastly larger and more well-known movement against revision of the U.S.-Japan Security Treaty (Anpo), Wesley Sasaki-Uemura shows how protests formed part of broader efforts to shape democratic practice in Japan.[60] In Kyoto, participants in very small-scale, local protests were doing something similar: while pushing to protect their interests, they were also working to develop and exercise methods of civic participation beyond the ballot box. These protests can be seen as part of a long process that began in the 1950s and evolved over the following decades into a vibrant culture of small-scale popular activism. Like anti-Anpo groups, communities fearing damage from the bullet train line used various activities to make their concerns heard: meetings and debates, petitions to the city assembly, visits to JNR's Osaka offices, dissemination of information, and sit-ins. These activities did not, for the most part, have a significant impact on the actual path of the line through their neighborhoods, but like the Tokyo protests that participants viewed from afar, whether intentionally or not, they formed part of "a struggle by ordinary Japanese to consolidate participatory rights in both state and society."[61] Using human networks and communication channels apart from the official framework of democratic action, organized at the local level around particular shared interests, these protests were early iterations of a new mode of participation in Japan, empowering marginalized constituencies and changing ideas of citizenship and political belonging.[62]

It is difficult to recapture the thoughts and feelings of the people in these communities, as they did not tend to leave behind detailed records. But the historian can catch a sideways glimpse through three types of sources, each with varying levels of reliability in providing this particular viewpoint. First, the city assembly meeting reports are, of course, created from the standpoint of the administration, but they contain many references to and sometimes summaries of both written and in-person petitions made by local groups regarding bullet train construction. These do not include the details of the complaints or reveal the inner thoughts of those making them, but they provide a sense of the quantity and general content of opposition statements. And they can be read "against the grain" of their bureaucratic production to reconstruct the implied critiques. Second, a lawyer who advised one group seeking compensation for eviction from their place of business, Nakabō Kōhei, described the process in his memoir. This was written decades later and, like the city assembly reports, from a different perspective than that of the protesters themselves, but it does provide a firsthand account and detailed narrative of one example of the numerous small-scale popular struggles ignited by bullet train construction.

A more questionable (but still valuable) source is a lightly fictionalized account of the protest movement, the 1966 novel *Shinkansen* by established writer and Kyoto city assembly member Nishiguchi Katsumi. This, too, represents an outsider's perspective, and its fictional genre means that no particular detail can be taken as fact. However, the author was directly involved in the political process, as a recently elected representative to the city assembly and a member of its JNR Policy Committee, and he aimed specifically to depict the heroic struggles and grassroots organization that formed around the new line. While the novel presents a glorified version of the story, viewed alongside more factual sources (including those just described, as well as newspaper reports of meetings and protests), it adds to our understanding of the activism surrounding the planning and construction of the tracks through Kyoto. In addition, it had a much wider circulation than the city assembly reports, which were filed away for the use of bureaucrats (and eventually historians). Though by no means a critical success, the novel was by a prominent local writer, and its easy availability five decades later (both as a stand-alone book and in the author's collected works) suggests a fairly wide circulation at the time.

The communities that would be disrupted by bullet train construction joined the process of decision making about the bullet train route quite late, after the bulk of the decisions had already been made. Their political marginalization was related to their lack of access to information (a source of inequality discussed in Chapter 3). Most did not know that the line would plow through their neighborhoods until JNR had already decided to run it through Kyoto and settled on a general path, at which point the agency began to hold information sessions for affected communities. The plan that began to solidify in early 1960 was exactly what city officials had repeatedly requested, passing through Kyoto along the south side of the existing line. However, local residents had not been privy to those requests, and public discussion of the potential route through the city prompted a wave of petitions, as new opposition movements began to coalesce in affected neighborhoods. But there was not much that people facing damage from the bullet train could do. JNR had a strong upper hand in terms of the legal structures of land acquisition, understanding of the processes involved in planning and construction, the political interests of local government representatives, and the symbolism of the project as a whole.

The legal framework was stacked against residents from the start, and their standing was further weakened during the early years of planning. The 1951 Land Expropriation Law allowed the government to compel the sale of land for use in the public interest. JNR officials were reluctant to resort to legal means of forced expropriation and had stated as much in both closed-door meetings and public pronouncements. When JNR president Sogō stated in January 1960 that the bullet train was likely to go through Kyoto, he noted a complaint from a local traditional craft association and expressed his desire to avoid use of the Land Expropriation Law.[63] That policy was reiterated two months later by the newly appointed JNR Trunk Line Department head. Visiting Kyoto to introduce himself to local officials and ask specifically for their cooperation with the process of land acquisition, he explained that JNR preferred to gain the understanding and consent of local people through discussion, rather than resorting to legal means.[64] But such public statements, reported in the newspaper, also underscored JNR's legal ability to force the sale of privately owned land. The potential consequences of resistance thus loomed in the background of all negotiations over land purchases for the line. Indeed, JNR officials sometimes encouraged cooperation by strategically reminding Kyoto's leaders of this possibility, a goal achieved even through statements that they hoped to avoid it.[65]

The people of Kyoto hardly needed such reminders. Recent experience had demonstrated the power of the national government in this regard, when the Construction Ministry made rare use of the Land Expropriation Law in January 1960 to overcome local resistance in order to complete the purchase of the last 20 square kilometers of land necessary for the Meishin Expressway. This included land with residential buildings and a shopping district in three Kyoto wards where JNR would need land for the bullet train: Minami, Fushimi, and Yamashina (Higashiyama). Authorization of expropriation under the law was made on January 13, 1960, and newspapers reported it the following day.[66] This coincided with JNR's announcement of the intermediate stations planned for the bullet train and the likely decision to have a bullet train stop in Kyoto alongside the existing station, which would more or less ensure that the tracks would run through Yamashina, directly east of the station. The ruling on the Meishin Expressway meant that the people of Kyoto were even more keenly aware than others of the potential for expropriation.[67] This context also exacerbated the sense of undue sacrifice that helped spur residents to political action. In fact, two wards where the Construction Ministry had turned to the Land Expropriation Law to obtain land for the expressway—Higashiyama and Fushimi—were among the first to form opposition committees and submit petitions regarding the bullet train route.[68]

The existing legal framework for expropriation of land involved a long and complicated multistep approval process, but the planning of several large and nationally significant infrastructure projects prompted lawmakers to craft an expedited system to be used in cases of particularly urgent public interest. A Construction Ministry committee began in-depth discussions on legislation to speed up the cumbersome bureaucratic process for forced land acquisition in August 1960, in the context of a need for speedy construction not only for the bullet train and new expressways but also for the many needs created by the upcoming 1964 Tokyo Olympics.[69] In August 1961, the Diet passed a new law, the Act on Special Measures Concerning Acquisition of Land for Public Use (*Kōkyō yōchi no shutoku ni kan suru tokubetsu sochi hō*), aimed at the "speed-up" of specially designated public projects, such as expressways and trunk railway lines.[70] The new law expanded eminent domain beyond the scope of the 1951 framework, allowing the government to acquire and immediately use land by paying estimated compensation to the existing owner. JNR's application for recognition of the New Tōkaidō Line as one such "Special Public Project" was approved on October 19, 1962. At that point, the agency had obtained the land needed for all but around 40 of the total 515 kilometers of the line.[71] Its inclusion in this new category of urgent public projects added to the tremendous pressure on the owners of those last remaining stretches of unpurchased land.

Beyond the legal framework, JNR was further empowered in negotiations by their control over information related to planning. Much of the official petitioning of JNR involved pleading for information about the route and station details, and the uncertainty surrounding these questions caused tremendous consternation among those who might be affected. In a November 1959 explanation session with JNR officials, one city assembly representative complained that with various proposals being floated, "Citizens are agitated. There are people going around saying all sorts of things. We just want a decision, so please tell us your plans."[72] Even when JNR announced its decision to run the line through the city with a stop at Kyoto Station, they promised to place it close to the Tōkaidō Main Line but included a wide swath of land in their preliminary surveys. This prompted a lively discussion in the city assembly. Representatives demanded that the details of the route be made clear, as the uncertainty of the broad survey area was "making local residents annoyingly uneasy."[73] Of course, JNR officials were not being secretive just for the sake of maintaining an upper hand in negotiations. Rather, they were trying to maintain some flexibility in order to find the optimal route. As the head of JNR's Osaka Trunk Line Construction Department explained to visiting Kyoto city assembly members, the process always involves some uncertainty. "There are cases in which you draw a line on a map, then survey the land. You put in the stakes, then put in a narrower line. And even after changing the site once and starting construction, you might move it another five or six meters."[74] He therefore begged for the officials' understanding that there might be unforeseen developments that force a change to tentative initial plans.

In response to persistent demands for information, JNR held a series of explanatory meetings with the general public in affected areas, including Minami, Sakyō, and Higashiyama Wards. The railway had initially resisted calls to hold such public meetings, arguing that this step would make more sense once they had drawn a fairly specific and certain map of the route to serve as a concrete basis for discussions.[75] But they gave in to pressure and held meetings before the parameters were clearly defined. Even then, railway representatives controlled the distribution of information about planning and compensation, while local residents relied on hearsay, rumor, and the selective information provided by JNR and city officials. As a result, the meetings simply resulted in greater frustration on the part of local residents and created an impression that JNR was being shifty and dodging their questions. Meeting with city assembly members, representatives of Minami Ward dismissed JNR's efforts at explanation, saying they had been vague and evasive without making any effort to understand the views of the local people.[76] Therefore, even after the meeting, because JNR failed to give adequate responses to their questions, residents continued to harbor strong concerns about whether tracks would be elevated and uneasiness about the prospects for fair compensation. The outcome was not the cooperation that JNR had sought but rather a plan for a strong, broadly supported movement to push for elevation of the bullet train, as well as the existing tracks, as the precondition for cooperation.

Nishiguchi's fictional portrayal colorfully represents this dynamic. Depicting a real-life planning meeting—which took place in February 1960 at the Rokusonnō Shrine, just west of Kyoto Station in Minami Ward[77]—Nishiguchi has JNR's head of land acquisition calm rising passions among the gathered residents and deflect their complaints and accusations with vague assurances of extensive compensation. But the story's omniscient narrator editorializes, "Never has there been such an empty promise that made fools of people."[78] With the benefit of hindsight, writing the novel after it was all over, Nishiguchi was able to see through JNR's obfuscations, something that was more difficult in the moment. Residents hungry for facts about something that would have a profound impact on their lives were reduced to dealing in rumors in ways that muddied their efforts to organize. For instance, later in the novel, we overhear a conversation among residents who had attended the Rokusonnō meeting, in which they speculate about toilets being emptied over their homes from the elevated tracks above. The novel includes a number of conversations peppered with statements beginning, "Rumor has it . . . ," followed by specious information about the route, the height of elevated tracks, the possibilities afforded by JNR's technological capabilities, and all manner of things.[79] Nishiguchi's novel thus publicized for a general audience what is palpable to the historian from the less widely accessible city assembly reports, namely that JNR's control of information helped them preclude, counter, and undercut public opposition.

JR's position was also strengthened by its monopoly on information what was possible. It was difficult for citizens' groups, or even Kyoto's local leadership, to press for an alternative to JNR's proposals when they could not be certain whether it was technologically or geologically feasible. Railway representatives often put off questioners and petitioners with vague statements about technological capabilities. This is not to say that they were lying but rather that their expertise gave them the ability to silence opposition. JNR had cited the limitations imposed by technological capabilities and land quality in their responses to local pressure from the start.[80] When JNR president Sogō spoke in Kyoto in January 1960, he promised that the agency would consider the demands of local residents, but added "within the limits of driving technology," an important and, because only JNR's engineers really knew the limits of that technology, quite flexible caveat.[81] JNR deployed arguments based on technological limitations in response to protests about the route, the problem of narrow spaces between the existing and new Tōkaidō tracks, and specific construction methods.[82] The same problem comes up repeatedly for petitioners in Nishiguchi's fictional rendition of the story; as they discuss possible alternatives to propose to JNR, they always come to a question they cannot answer: Is it technologically possible?[83]

Residents were further hindered by the failure of local government institutions to work on their behalf. Assembly reports show that affected communities' interests were being unevenly (and for the most part poorly) served by the basic infrastructures of democratic representation—prefectural and city assemblies, whose members were elected by each district, as well as local representation in the Diet—because residents' interests diverged from or directly conflicted with the concerns of more powerful groups and individuals. It is not that their representatives remained completely silent. For instance, some city assembly members raised questions about potential problems stemming from the need for evictions when the chair of the JNR Policy Committee called for cooperation with the campaign to attract the bullet train.[84] But, like many large construction projects, this one presented a problem of conflict of scale: what is useful or beneficial on one scale (the nation, the city) is harmful on another (specific neighborhoods, individual households). Representatives sometimes tried to promote the interests of their constituents, but for the most part, they succumbed to the dominant forces at the larger scale that propelled the project forward.

Residents certainly voiced their opinions: petitions from commuties fearing eviction started to flow into the assembly almost immediafter JNR's informal decision in November 1959 to have a stop in the vicinity of Kyoto. Several resident groups presented petitions in person and in writing over the following months, and this trickle became a flood after the route was formally announced in March 1960, as new opposition groups coalesced.[85] In the first half of 1960, the assembly received at least thirty separate petitions regarding the bullet train route from both existing neighborhood associations and new groups formed for the specific purpose of opposing the line.[86]

The details of the numerous petitions submitted to the city assembly in the months after JNR's decision are mostly lost, but based on the brief descriptions preserved in the city assembly reports, they can be divided roughly into three categories. Many simply expressed opposition to having the tracks plow through their neighborhoods. Others, from neighborhoods in the city's southwestern sector, called for a specific change to the route. And later, once the route had been finalized, the focus of petitions turned to the practical but highly contentious and emotional issues of eviction and compensation. The city assembly, as a body, paid little attention to these many petitions, and only the specific request for a shift in the line gained the support of the city government, most likely because it was in accord with an existing plan for industrial development in the affected districts. In fact, that request was initially raised within the city assembly by the representative of Fushimi Ward, which sprawls horizontally across the southern part of Kyoto, its western border just reaching the bullet train tracks on their way out of the city toward Osaka. In response to the March 9, 1960, announcement of the route, the Fushimi representative complained, "Fushimi Ward, which is already crossed by the Meishin Expressway, is again being cut across by the Shinkansen. This will hinder the development of Fushimi."[87] Opposition groups coalesced in the area, including an alliance among six companies with factories in the Kuze district (which extends into Minami Ward), and presented petitions arguing that, if the tracks followed the path JNR proposed, they would form an obstacle to an ongoing plan for industrial development. These petitions promised cooperation with construction but asked for a slight westward shift to the route, so that it would not pass through the middle of a planned "industrial belt," which included the construction of about twenty new factories.[88] This case demonstrates the importance to local groups of support by the city administration, as the head of JNR's Trunk Line Department agreed that JNR would be willing to move the line on the basis of the assembly's recommendation.[89] A representative of the mayor's office stated explicitly that negotiations with JNR would proceed on the basis of city planning, not the objections of residents, and the city assembly agreed with that approach.[90]

The nearby Kuzetonoshirochō district of Minami Ward opposed this section of the route in even starker terms, calling it "a matter of life or death for us," and the people of the ward gathered on March 12 to start a movement to change the route, forming the Union Resolved to Oppose the JNR Shinkansen. However, they did not have the backing of industrial interests. Rather, their opposition was based on the prediction that having the bullet train pass through their district "would destroy about half of the homes and farmland."[91] Today, the bullet train does, in fact, slice through the middle of the district.

The Kuze-Fushimi industrial belt was an exceptional case. Most petitioners found little or no support from the city assembly, whose priority was having the bullet train stop at the existing Kyoto Station, a conflict with the interests of communities along the proposed route. The local administrative body's unwillingness to risk losing a stop on the line by supporting affected communities was made clear at their March 21, 1960, meeting (soon after JNR's announcement of the planned route through the city and the subsequent outpouring of petitions in opposition). After hearing yet another petition and discussing future plans for continued pressure on JNR regarding elevated tracks for both the new and existing lines, the JNR Policy Committee asked each political party group for their official position on the route. The responses differed in some minor details, but they reflected unanimous support for JNR's plan. The conservative LDP members started off with a call to support the railway agency, while pushing for elevated tracks. Their spokesperson also advocated protecting the emerging western industrial belt by pressing for the line to run along the route demanded by local groups. The Socialist Party followed, echoing the LDP on the importance of elevating the tracks of both the new and existing lines. Regarding "other aspects," they proposed that the assembly "not even touch it" (*issai tatchi shinai*) and instead simply pass along petitions made by local groups to JNR. Similar responses were forthcoming from both the Citizens' Party and the

Communist Party, the latter adding a demand that JNR consider city planning in deciding the route. The Democratic Socialist Party spokesperson addressed the issue of petitions most directly, stating that it would be sufficient to communicate the substance of opposition to JNR and asking, "At this point, should we throw ourselves on the fire?"[92] It is clear from these statements that the members of the city assembly prioritized these particular goals to the extent that pushing for anything else was seen as potentially detracting from those goals, the political equivalent of throwing themselves on the fire.

Novelist-politician Nishiguchi took a jaundiced view of local leaders and hinted at nefarious realities behind the administrative support for JNR in his novel about the bullet train. The narrator describes one local figure, the head of the Hachijō Commerce Association, "playing a role, crying about forced eviction, while planning to get a high price for his real estate," and simultaneously engaging in backroom negotiations with JNR's land acquisition officials, based on the principle that "he who hesitates is lost, and the important thing is to be resourceful."[93] He and a colleague plot to get the poor people who live along the track riled up in the hopes that popular opposition will further raise the price for their own land. Later in the novel, during a public discussion of the issue, a local landowner named Kotake, who is a member of the city assembly with the conservative party (dubbed the Jizaitō or "free action" party in the novel), calls for cooperation with JNR on this "great national project" and a conciliatory attitude in negotiations. The narrator notes that in the past, people would have accepted this, but a new activist spirit invigorating the community prompted an uproar in response, with accusatory shouts that that Kotake just wants to make sure the bullet train comes through the neighborhood so that he can unload some worthless land for a high price.[94]

It is notable that this fictional character's accusation is almost a direct quote from an incident in nearby Shiga Prefecture, which had been reported in the major daily *Yomiuri shinbun*. A Buddhist temple in the bullet train's path was refusing to sell 33 square meters of land where its sutra repository was located, so JNR took measures to force expropriation of the land and asked the village's deputy mayor to negotiate. The temple responded with an accusation that the deputy mayor was taking JNR's side so as to fetch a high price for his own land.[95] Such local disputes rarely made the news; this conflict became a national story primarily because it was one of the few cases in which JNR engaged in forced expropriation of land. Presumably, most such incidents passed under the radar of the broader public, highlighting the importance of looking beyond the official record for information. The accusation echoes the suspicion that landowners' resistance to sell was sometimes just a ploy to get a higher price.[96] In addition, the appearance of a number of similar scenarios in popular culture suggests the likelihood that it was not uncommon.[97] Such stories must be read with skepticism about their adherence to the facts, but they nevertheless capture some historical truth.

In spite of these divided interests, the city assembly did press to some extent for changes to benefit their negatively impacted constituents. For instance, a May 1960 joint statement to JNR from the city and prefectural assemblies and the chamber of commerce focused on the usual demands regarding the location and elevation of the tracks, but it also added a request that "facilities constructed through national planning do not destroy the environment supporting the lifestyle of people living along tracks or hinder integrated urban development, but rather assist and improve them."[98] The statement then specified nine areas of concern, in most cases suggesting that problems could be avoided through the use of elevated track. This included the already familiar requests regarding elevation of the tracks of the Tōkaidō Main Line and placement of the new line in the industrial zone, but also drew attention to the Hachijō neighborhood of residences and shops on the south side of the station and the area west of the station to Rokusonnō Shrine, two sites of intense petitioning from local groups. The assembly later helped to mediate negotiations between JNR and local groups in the most intractable disputes over land acquisition and compensation.[99] But for the most part, they took a passive stance and let their most vulnerable, poor, and powerless citizens fend for themselves against the central government agency.

Given the legal context and conflicting interests with the city administration, affected communities did not have many options for pressing their cause. Frustrated at their lack of power using existing democratic infrastructures and desperate to safeguard their homes and livelihoods, local groups formed alliances with individuals who had specialized knowledge or skills to help them forge new paths for communicating and promoting

their interests, including obtaining accurate information, changi
route of the tracks to lessen the damage to their districts, or receiv
equate compensation for their land or forced eviction from rented ᵤₚ⁻⁻
These spontaneous coalitions, self-generated in pursuit of a shared interest, formed an alternative outlet for expression of their views.[100] They gathered for debate and discussion, wrote up petitions to local leaders, distributed flyers, and held sit-ins to obstruct construction. Engaging in such
activities, they themselves formed a kind of human infrastructure of democracy. Describing what he calls "people as infrastructure," AbdouMaliq
Simone argues that economic collaborations among residents of marginalized communities in the inner city of Johannesburg conjoin to form a complex social infrastructure, in which every instance of collaboration creates
a connection that can be relied upon in the future.[101] Construction of such
"invisible" or "soft" infrastructure was part of the official policies of the
prefectural Ninagawa administration, which proclaimed the principles of
"protecting the life and livelihood" of citizens and "invisible construction"
of economic, social, and cultural human relationships that would allow local citizens to reap the benefits of centralized "visible construction," like
the bullet train.[102] In that sense, the people of Hachijō, Kuzetonoshirochō,
and neighborhoods all along the line through the city were engaged in their
own infrastructural project, building new connections even as they were
being physically displaced. Those connections, though tentative and pre-
carious, like the human infrastructures Simone describes, helped them in
their long struggle to gain fair compensation from the government.

The result was a shifting landscape of overlapping groups that coalesced, divided, and reconstituted in different forms on the basis of shared
interests and differing views of how to respond to the threat posed to local communities. Some opposition used existing frameworks for collaboration, such as industry and neighborhood associations. For instance, the
Imagumano Hiyoshi Pottery Manufacturers' Association organized the
writing and submission of demands to the city assembly, and several neighborhood associations submitted petitions.[103] But many new groups formed
out of shared concerns, such as the JNR Shinkansen Site Community Association (*Kokutetsu Shinkansen yōchi jichikai*), inaugurated in November 1961 by people with connection to land that would be necessary for the
bullet train route; the JNR Shinkansen Yamashina Opposition Committee

(*Kokutetsu Shinkansen Yamashina hantai iinkai*); the JNR Policy Liaison Council (*Kokutetsu taisaku renraku kyōgikai*), jointly formed in March 1960 by residents of Yamashina, Imagumano, Hachijō, and Mukōmachi—all neighborhoods through which the bullet train currently runs; and the Kyoto Minami Ward Eviction Opposition Right to Living Protection Alliance (*Kyoto Minami-ku Shinkansen tachinoki hantai seikatsuken yōgo dōmei*), among others.[104]

As it became clear that JNR would not budge on the line's location around the station area, petitioners shifted their focus from trying to prevent the devastation of their neighborhoods to ensuring that they would receive adequate compensation. Many of the petitions to the city assembly were related to assistance or compensation for two categories of people not directly included in JNR's compensation scheme: those who rented residential or business space and, therefore, would be evicted without receiving the compensation that was due to landowners; and those who were not evicted (and, therefore, not compensated) but nonetheless suffered from bullet train construction because they ended up squeezed into what came to be known as "pocket areas" (*poketto chitai*), the narrow spaces between existing tracks and the new line just a short block away. Some proposals included a last-ditch plea that the bullet train line hew more closely to the existing line, so as to eliminate pocket areas, but most petitioners were not pressing for any change to the route or trying to stop the line from coming through their neighborhoods. Rather, they simply sought fair compensation for the disruption it caused to their lives and livelihoods.

In November 1961, having relented after nearly two years of resisting the line by refusing to allow JNR to conduct surveys on their land, the JNR Policy Kyoto Districts Association (*Kokutetsu taisaku Kyoto chiku rengōkai*)—comprising communities all along the line from Higashiyama on the eastern side of the city to Karahashi in Minami Ward (west of the station, just before the track curves southward)—sent about twenty representatives to present a vehement appeal at a meeting of the city assembly's JNR Policy Committee, with the mayor and deputy mayor both in attendance. The association chair stated, "Among people being evicted, many are renters and have no place to go, so they are in difficulty. We are not opposed to construction of the Shinkansen, but rather we are asking the city and the assembly to cooperate in ensuring that the local people do not have to bear an unfair sacrifice."[105] Their demands were not to stop construction but simply that the city help the people affected in four specific ways: provide a substitute lot to those evicted, guarantee housing for renters and apartment dwellers, purchase the geographical pocket wedged between the existing line and the new line, and implement a policy to assist the businesses on the shopping street currently south of the station.[106]

Mayor Takayama responded to these demands with the explanation that finances would not allow the purchase of pocket areas, pointing out that the expense of it would reduce the government's capacity in every other area. Promising to work for the benefit of the citizens, he noted that the city was investigating the possibility of building public housing and making use of the land beneath elevated tracks.[107] Here, too, citizens came up against the hard realities of city governance, which had to take into account many competing interests; their concerns were far below the top priority. Even when they did receive assistance or compensation for their losses, they had to fight at every incremental step for what they referred to as their "right to living" (seikatsuken), which is encoded in Article 25 of the Japanese Constitution as "the right to maintain the minimum standards of wholesome and cultured living."[108] For instance, though the city did ultimately build new public housing for evictees, even after moving in, they had to continue to struggle against JNR, which tried to recoup its share of costs by collecting concession fees for the new apartments from the eviction compensation funds.[109] As Simon Avenell demonstrates, this idea of "daily life" was central to the development of citizen activism in the immediate postwar decades. The local groups grappling with JNR or city leaders are an example of sociologist Katō Hidetoshi's 1960 observation that ordinary people joined political movements not usually out of some lofty moral awakening but instead out of the rather self-centered idea that "in order to protect their 'lives' [seikatsu] political action was necessary."[110] In such ways, local concerns fueled and shaped popular movements and thus contributed to a reimagination of citizenship in terms of ordinary lives taking place in local spaces, as opposed to a national-scale, state-centered notion. Thus, the "local egoism" of movements to defend or promote the interests of a particular community became part of a nationwide revision of the concept of citizenship and the balance of the individual's rights and responsibilities against the interests of the state and the nation.[111]

One path for local communities whose needs were not being met by either the city administration or JNR was to work outside the official democratic system and take up alternative methods of political expression in order to make their voices heard. Like those who refused to allow JNR access to complete on-site land surveys, they employed methods that might be categorized, borrowing James Scott's term, as "weapons of the weak" to force JNR and the city to accord them fair treatment. As Scott explains, "Everyday forms of resistance make no headlines," but their agglomeration can create real obstacles for the state and, in such ways, impel policy changes.[112] Resistance by local groups in Kyoto were mostly small-scale, low-impact efforts, barely visible in the historical record. But we can see their actions through the eyes of those from whom they sought help and expertise. One of the biggest problems for JNR was buying land, as many people resisted selling, whether they were holding out for a higher price or actually intent on staying in place. But, as the petitions discussed earlier suggest, those who were renting space for their homes or businesses could be left out of frameworks for compensation. One group of shopkeepers being evicted from their rented space, feeling they had unfairly been denied compensation, turned to a local lawyer, Nakabō Kōhei, who encouraged them to engage in a sit-in campaign. This small protest not only was successful in gaining them fair compensation but also helped spread a growing mode of popular political action.

Nakabō later wrote about their case in his memoir. The group consisted of shopkeepers who rented space to sell fish, tea, flowers, tempura, tofu, textiles, and other merchandise in what Nakabō refers to only as "M building," located about 500 meters from the station on land where the bullet train's elevated track now runs through. When the shopkeepers learned that JNR was buying the building with plans for demolition, they petitioned the agency for compensation but were put off with a promise that the issue would be discussed once negotiations with the building owners were completed. The owners themselves threatened the shopkeepers that they would lose their security deposits if they interfered. Almost a year and a half after the April 1961 sale of the building, the shopkeepers still had not received any word of compensation, much less actual funds. The group went in person to the Osaka Shinkansen Construction Department in September 1962, but that office washed its hands of the situation, claiming they had already

bought and paid a large sum for the building and land, and the shopkeepers should look to the former owners for compensation. When the former
owners did not respond to their requests, they declared that they would not
cooperate with construction, and JNR responded with an offer of compen-
sation that Nakabō calls "absolutely unacceptable."[113] → *Civilian protest was not taken seriously at first*

Nakabō, who would later make his name in representing victims of the
infamous Morinaga milk poisoning incident, looked into their case and
concluded that there was no legal basis for opposition. Neither JNR nor
the building's owners had actually violated the law. Any normal lawyer, he
wrote, would give up at that point. But public perceptions of JNR had recently taken a hit with news that the head of land purchasing in the Kyoto
area had been arrested for corruption. In that context, Nakabō felt a strong
sense of JNR as the guilty party and wanted to help protect the rights of
leaseholders injured by such cases of collusion.[114] For Nakabō, this case
presented an opportunity to make democracy work for "the people" against
powerful interests. His anger at the situation led him to set aside his legal
training and forge a path into what was, for him as for the shopkeepers he
was advising, completely unfamiliar ground. Instead of giving up, he suggested that the shopkeepers hold a sit-in in order to mobilize public opinion. With no experience or knowledge about protests, he was acting "on in-
stinct." However, he writes, "by borrowing the strength of public opinion,
even the weaker party has a chance of winning."[115]

When construction started, the shopkeepers and their families, including small children carried on their backs, rushed to sit in and distributed
flyers to explain their actions to the public. Because of this printed explanation, which blamed the building owners for colluding with JNR and treat-
ing them unfairly, the owners accused them of libel. The fear of such legal action contributed to the invisibility of such protests—in fact, Nakabō
nearly counseled his clients to abandon their plans to distribute the flyer
because he thought it would provoke exactly such a response. But contrary to his expectation that they would be cowed by the accusation, they
bravely faced the possibility of arrest and continued with their sit-in. Ultimately, not a single person was arrested, and JNR paid out the desired
compensation.[116] Though this method gave voice to marginalized groups,
it differs in one important aspect from Scott's explanation of "the weapons of the weak" as hard-to-detect actions that do not bring about direct

confrontation with authority.[117] Far from escaping punishment by avoiding detection, the sit-in was meant to gain the widest possible attention in order to put the pressure of public opinion on powerful individuals and institutions. Having tried everything, the desperate shopkeepers were ready to risk arrest to protect their rights. At the end of a long career spent fighting for underdogs against powerful institutions, Nakabō wrote about his experience with the shopkeepers of M building pressing for their rights against JNR as the origin of that trajectory. Working outside the regular legal infrastructure was, for him and for the protesters seeking fair compensation, a way to make democracy real and to make it work for everyone, even the least powerful. In this way, the protests surrounding the bullet train were one of the many early contributors to the trend toward issue-specific citizens' movements that was emerging at the time and took on strength over subsequent decades.

Another local community sought help from a sympathetic city assembly member who used his knowledge of local politics and his familiarity with the bullet train plans through participation in the assembly's JNR Policy Committee to help them pursue their goals in different ways. We can read a semifictional version of their story in Nishiguchi's novel. The author grew up in Kyoto's Fushimi-chō, in a neighborhood depicted in his 1956 novel *Red-Light District* (*Kuruwa*). He was already a successful novelist by the time the struggle over the bullet train came to his city: that earlier novel not only became a best-seller but was also nominated for the Naoki Prize, one of Japan's most prestigious literary prizes, awarded semiannually to the best work in popular literature. In addition to his literary activities, Nishiguchi was active in local politics as a member of the Communist Party from 1946. He was elected to the Kyoto City Assembly in 1959 and joined the JNR Policy Committee. In that position, he participated in the efforts to press for a bullet train stop in Kyoto, heard petitions from local groups, and met with protesters.

Nishiguchi wrote his novel about the local bullet train disputes in Kyoto as a record of one instance of people standing up for their rights, in order to provide for public use a template for democratic action against powerful institutional forces. The result is a barely fictionalized version of the story; many of the events depicted are verifiable in the historical record, though names are changed and private conversations were probably invented by the author. The novel details efforts of local groups to push back against JNR's plan to demolish their neighborhoods and condemn remaining homes and businesses to vibrations, dust, and noise, both during construction and thereafter from constantly passing trains. In his preface to the novel, he loosely paraphrases Immanuel Kant, writing, "If people resign themselves to being worms, they will probably be squashed just like a worm."[118] When Kant writes, in *Metaphysics of Morals*, that "one who makes himself a worm cannot complain afterwards if people step on him," he is talking about the importance of maintaining and enacting self-respect.[119] Nishiguchi's point in making this reference is that even people with little power can defend their rights against dominant forces if they rely on their own self-respect to stand up for those rights, thereby commanding the respect of others.

In his preface to the first edition of the novel, Nishiguchi claims rather disingenuously, no doubt in order to preclude accusations of libel, that "this work is a product of my imagination depicting the resistance to the coercive construction of the Tōkaidō Shinkansen by poor, unknown people. If anything that resembles what is in the work actually happened in the past, it is just because that resembles what is in my imagination."[120] Given the strength of libel law in Japan, it was not uncommon for authors depicting recent events to protect themselves with such assertions. But many years later, Nishiguchi referred to the work as a "semi-document of the battle against JNR authorities by local residents and the local assembly."[121] Indeed, the fictionalization is quite thin. The story centers on a historically important midsized city called "Q-to" (written with the Roman letter Q followed by the second character of "Kyoto"). It begins with an explanatory meeting between JNR officials and local citizens at a shrine called Rikusonnō, which lies directly in the path of the new line. The visiting JNR official is told by his assistant before the meeting of the building's historic importance, as it enshrines the founder of the Kamakura shogunate, Minamoto no Yoritomo (1147–1199).[122] The real-life counterpart of this meeting, as noted earlier, took place in February 1960 at the Rokusonnō Shrine, built to house the soul of Minamoto no Tsunetomo (894–961), the progenitor of the Seiwa Minamoto warrior clan and Yoritomo's forebear by several generations. By making such minor alterations of fact, Nishiguchi puts a veneer of fiction—and plausible deniability within libel law—on what is fundamentally a true story.

Nishiguchi's stance on the struggle between local communities and JNR is clear. He gives JNR's head of land acquisition for the project the name Minchi Shimeo, a made-up name created from characters that can be translated as "People's-Land Taker-Man." Aside from the meaning inherent in his strange name, Minchi's attitude had already been made clear to the reader by his annoyed response upon learning of the shrine's cultural significance: "Q-to has nuisances rolling out to the outskirts of the station area. But fine, it won't cost much to buy out a poor shrine."[123] Even Kyoto's historic spaces were not, for Nishiguchi's fictionalized JNR authorities, important sites to be preserved but obstacles that will increase construction costs. At the meeting, the presentations by both JNR and local officials strike a familiar chord. Minchi, dressed in a stylish suit with a scented pocket square, "presents contents he had repeated many times over like a rotary press," explaining that Japan's wonderful economic growth had created a transportation bottleneck, meaning that people will not be able to get fresh food, and Japan will devolve into turmoil resembling that of the immediate postwar years, but that the problem could be solved by the "dream super-express," which would run faster than any train in the world. He then added that the course had been changed in response to the passionate petitioning from Q-to leaders, who argued that "of course, this line should pass through Q-to, which is world-famous as the thousand-year-old capital."[124]

The combination of emotions evoked by Minchi's speech—painful recollections of postwar hardships, followed up by nostalgia for childhood fascination with steam trains and local pride in the city's cultural importance—initially wins the crowd over, and there are murmurs of excitement around the room. But gradually, doubts begin to emerge, especially after the officials display a map showing the general path the line will take, a wide red swath through an area of dense housing south of the station. Local residents are shocked to hear for the first time of the plan for the bullet train track to come right through their neighborhoods. The Q-to city head of public works jumps in with an explanation that sounds, for readers, another familiar note, arguing that the placement of the station is the most rational choice, given the city's position "as a global cultural tourism city," because it will be the most convenient for tourists.[125] Frustrated at his own inability to promote the interests of his constituency against the powerful forces aligned behind the bullet train project, Nishiguchi echoes these standard claims about the new line's significance to Kyoto in order to emphasize its irrelevance to the people whose lives would be most disrupted by it.

The community's realization that they might all lose their homes for the sake of this national project leads to a discussion of how democracy works. To each accusation of undemocratic behavior, Minchi responds with a different definition of democratic processes. For instance, a local grocer, Daikon Genji, described as glaring at Minchi like a Niō guardian king, expressed doubt that it was really in response to Q-to's wishes that the line was changed to come through the city, given that the people affected had not heard any thing about it: "Is it democratic for big shots alone to selfishly decide without consulting with the most important parties, us local people?"[126] Minchi responds by pointing out that JNR had come to the neighborhood on this occasion for the very purpose of gaining the people's consent through democratic discussion. Gradually, the gathered citizens begin to feel that their own representatives had conspired against them, making decisions about urban planning at their expense without public discussion. As this realization takes hold, a shock of anger reverberates around the room. To deflect their blame, Minchi argues that this is not the fault of JNR but rather an internal Q-to problem, pointing out that because Japan has a democratic system of government, the decisions of the elected city and prefectural assemblies must be taken as the people's opinion.[127]

This exchange is significant not only for its content but also for the styles of speech depicted. Minchi speaks in standard Japanese with textbook grammar in what the narrator describes as an effeminate voice. The local residents, on the other hand, all speak with a strong Kansai dialect. This not only marks their local specificity but also signals another aspect of their distinct identities and positions in the process of bullet train planning. As Michael Cronin argues regarding literary representations of modern Osaka, "More than any narrative description of local places, [or] customs, . . . the written approximation of characters' local speech produces the idea of Osaka in literature," which is consistently figured as "treasonous," resisting the central authority emanating from Tokyo.[128] In this sense, Nishiguchi's consistent use of the Kansai dialect to represent the speech of local residents of the station area can be seen as a representation of their resistance. With this scene, Nishiguchi suggests that the rhetoric of "great

Non-democratic

national project" used to promote the development of high-speed rail pa-
pered over the fundamental inequalities that continued to plague Japanese
society even two decades after the establishment of a democratic system.
But he also hints at the resilience of a people for whom resistance against
central authority is part of their local identity.

Nishiguchi conveys his message at another level by representing the activities of the city assembly. The author depicts himself in his novel as a
newly elected city assembly member named Saijō Bunpei, a representative
who has just entered into a system he soon finds to be rigged in favor of
the powerful, and who eventually tries to work outside that system to help
marginalized communities. We meet Saijō on an overnight train to Tokyo,
on his first of many pilgrimages to the centers of bureaucratic and political power to plead the city's case. Proceeding fruitlessly from one office to
the next at JNR headquarters, he observes his colleagues having the same
meaningless exchange with each bureaucrat they meet. The Kyoto representatives present their demands, and each JNR official mechanically gives
the identical response that he will convey their message to his superiors.
Saijō gradually begins to feel that he is in "a modern *sankin kōtai* drama,"
referring to the Tokugawa-era system of alternate attendance, which required regional lords to spend every other year in the capital—a political
performance with no real impact that, nonetheless, cannot be neglected.
The petitions themselves had no practical effect, but "if you do not do it,
then rather than zero, it becomes a minus," while the success of the petition is decided by secret behind-the-scenes deals—another kind of invisible infrastructure.[129]

Frustrated with this process, Saijō finally speaks up for the first time,
interjecting a sarcastic comment about the uselessness of these conven-
tional responses and a warning of the terrible opposition movement that
will arise if JNR chooses a route that ignores the sacrifices required of local communities and the damage they will suffer. This blows up into an
argument in which the conservative party committee member concludes,
"That's why I hate Reds [*Aka*]."[130] The tensions in Japanese society between the political left and right underpin this story throughout. It is set
against the backdrop of the massive anti-Anpo demonstrations, and several characters try to connect their protests against the bullet train to larger
movements, including the Anpo protests and a reinvigorated labor movement. From the opposite perspective, characters frequently refer to anyone

opposing official plans as "Reds" and question the trustworthiness of others by noting their connections to the Communist Party.[131]

In the novel, as in real life, groups of people come together by forming alliances based on shared interests in order to gain strength through numbers in facing JNR. People meet to discuss ideas, goals, and strategies in the context of block association meetings or more informally in the back rooms of their businesses or around the well by their barrack-style housing. Groups form, splinter, then realign or combine forces with other communities. Nishiguchi shows the formation of a group that adopts the name "Alliance to Oppose Minami Ward Shinkansen Evictions" and depicts a meeting (again in "Rikusonnō" Shrine) between this group and five city assembly representatives. The discussion becomes heated as people criticize their representatives for failing to protect them. One of those representatives finally turns the tables, assuring the gathered crowd that the assembly is working hard for them and blaming the people themselves for the "ruckus" of the day's meeting. Representative Uramichi (whose name is a homophone for a word meaning "unfair methods"), though a member of the Progressive Party, has spent the meeting defending JNR's plan against the people's complaints, which he labels "irresponsible thinking." He concludes the meeting with a call for creation of an alternative organization: "There was a ruckus today, but we can avoid repeating it if you all form a negotiating group appropriate to sensible people. Some of us have other business, so we have to go now, but we expect that you will soon establish a new local people's organization with which we can discuss things closely and frankly." This has the effect of splitting the existing group, as some agree that the best course of action is to make a new, more geographically inclusive group, while others vow to double down, promising to cooperate only "if the demands you decide upon match some of ours."[132]

Nishiguchi hoped that by demonstrating the rocky process of building strength through unity, his novel would become a "manual" for democratic action. In the afterword to a 1973 reprinting of the novel, he wrote that the struggle with JNR continued for several years, and the rail agency finally "retreated in the face of the people's tenacious resistance and our investigations just before the exposure of Shinkansen 'corruption.'" This experience, he continued, "was a keen lesson that democracy is not a theory, but something to fight for tenaciously."[133] The people of Kyoto had gained a new understanding of the workings of democracy, and Nishiguchi hoped to share their hard-won knowledge with a national audience that increasingly needed it in the context of expanding infrastructural development. He continued with a clear reference to the best-selling book *Nihon rettō kaizōron* ("Plan to reconstruct the Japanese archipelago"), published the previous year under the name of Tanaka Kakuei (by then prime minister), a plan for national development that called for bullet trains traversing the entire nation, as well as highways and other national construction projects: "Now, at a time when, under the command to 'reconstruct' the Japanese archipelago (*Nihon rettō 'kaizō'*), the Tōhoku, Jōetsu, San'yō, and Kyūshū bullet trains are being pushed forward by the sacrifices of the people of each region, I will be happy if this novel can become a reference not only for those people, but generally for all people unfairly forced into eviction by so-called public projects. . . . That is the reason I consented to republish it, in spite of its immaturity in terms of artistic merit."[134]

Nishiguchi's appraisal of his own work was clear-eyed and honest: the novel is really not very good. It did not receive nearly the acclaim (or, seemingly, the sales) of his earlier work. We cannot know whether anyone used the novel, as he hoped they would, as a how-to manual for protesting forced eviction for future railway and other infrastructure projects. But the activism of Kyoto's least powerful citizens was part of the beginning of a new mode of political action in Japan. On a local scale, Ellis Krauss has shown the political impact of a new demand for participation by ordinary citizens in local planning and decision making in the mid-1960s, which often focused on quality-of-life concerns, like pollution and other damages from industry or central government construction projects. Governor Ninagawa, whose administration actively supported such spontaneous movements, benefited at the ballot box from a new sense of the rights and responsibilities of citizens and the obligation of local administrations to be responsive to citizen input.[135] More broadly, protests surrounded the subsequent expansion of the bullet train system, not only by people who resisted eviction but also by groups concerned about noise pollution and other damage to quality of life along the tracks.

The three-way tug-of-war between JNR, Kyoto's city leadership, and local communities affected perceptions of the bullet train, metropolitan identity, and informal political infrastructures. The line's very presence cutting through the city and the changes that presence has wrought have naturalized its path, such that it now seems impossible that it would have bypassed the ancient capital. And the notion of Kyoto as an international culture tourism city seems to go without saying, as throngs of foreign visitors regularly arrive via high-speed rail from Tokyo to see the city's famous temples, gardens, and palaces. Similarly, citizen protest soon became an essential and lasting aspect of city planning and urban development in Kyoto—for instance, an ongoing cooperative movement to clean up the Kamo River and beautify the surrounding area that began in 1964—and eventually a normal part of the political process across the nation.[136] This story of competing interests highlights the contingency of those outcomes and the important function of perceptions of the bullet train and anticipation of its potential impact in guiding the actions of disparate groups. While JNR's planners emphasized the technical functions of high-speed rail in determining its path, Kyoto's leadership made a conscious and concerted effort to divert some of the bullet train's aesthetic force toward their own urban planning endeavors. Protesters perceived that force in a very different way, less sleek modernity than both source and symbol of the destruction of their communities and their own powerlessness to stop it.

Though nowhere else did resistance cause as much trouble for JNR as in Kyoto, protests all along the line influenced the meanings of high-speed rail on a wider geographical scope. Planners, pundits, and producers of popular culture used the new mobilities afforded by the line to reimagine the space of the Tōkaidō region, while residents and passengers reconsidered their own positions within it. This included anticipation (and fear) of new ways of using local spaces in the cities being tied more closely together by the line and of changes to relationships among the peoples who lived there. The following chapter will consider how divisions surrounding the significance of the bullet train that took place in Kyoto during planning and construction were echoed along the length of the line, from Tokyo to Osaka. Just as Kyoto's residents challenged JNR's sunny rhetoric of the new line benefiting the nation, broader popular discourse reflected overlapping and contradictory views of its impact on the spaces it traversed. Chapter 2 shows how, while the line physically reshaped the cities where it stopped, the very possibility of a day trip changed residents' views not only of individual cities but also of their relation to each other and their position within the Tōkaidō region and the nation as a whole, as well as reconsidering the very nature of the coastal area along the tracks.

2 Reconstructing the Tōkaidō

A TRAIN RACING ALONG a fixed line, connecting distant cities but also dividing the landscape, redefines the space it slices through. As the case of Kyoto Station demonstrates, railroads are tools wielded by bureaucrats, elected officials, and other planners for the reorganization of urban and regional space, but they are also forces that can damage communities and destroy natural and built landscapes in their paths. Whether beneficial or harmful, railway infrastructure reshapes the ways people use and envision those spaces. Zooming out from a single city to take in the full span of the Tokyo-Osaka route shows how the potential impact on local identity that Kyoto's leaders sought to capture in their campaign for a bullet train station affected the region as a whole, including the terminal cities and the relationship between them. Questions of politics, culture, mobility, and urban development come together in considering the impact of the bullet train on the space of the Tōkaidō region over several years of planning, construction, and eventual operation.

Emerging at the intersection of rapid economic growth, new approaches to city planning, and the steady urbanization of Japan, the New Tōkaidō Line played a significant role in reshaping the urban geographies of Tokyo, Osaka, and the spaces in between. A consideration of the bullet train in terms of the social construction of space—the ways in which people think about and use physical spaces—demonstrates the interaction between ideas and the built environment.[1] The public discussions of the bullet train before and after its debut show the mutual influence between the contours of this space and how people not only lived in but also envisioned and understood it. The construction of the new line did not have a profound effect on the physical landscape of Japan. Though the tracks veered in some areas from the path of the existing Tōkaidō line and ran through a few new stations, the line mostly transported people where they had already been going in large numbers. But it had an outsized impact on the spatial structure of the Tōkaidō region if this space is considered in not only physical but also conceptual terms.

The new possibility of traveling between Japan's two largest cities in only three hours on the line's fastest Hikari super-express changed both patterns of mobility and the popular imagination of the space of the Tōkaidō. It created novel spaces and inspired new ideas about the cities connected by the train and their residents. For Tokyo, in particular, the new line was assigned a role in plans for purposeful urban change, in contrast to the haphazard rebuilding that took place in the desperate periods of recovery after both the 1923 earthquake and the World War II firebombings that devastated the capital. In fact, the urban reimaginings of the 1960s were based in part on the understanding that the unguided reconstruction of earlier years had resulted in a flawed city that would soon be paralyzed by its own growth. This view of urban development created a perceived need to rethink some of the fundamental structures of Tokyo and Japan's other major cities, as well as the connections between them. On the other hand, it also brought into high relief the conflict between the goals of industrial and economic development and cultural and environmental preservation, a fundamental schism emerging in the context of accelerating economic growth and expanding urban sprawl.

The basic shape of the line, once settled, was durable. But the social reimagination of the space continued even after it was built, through the theories of urban planners, the combined movements of millions of people, representations of the bullet train in cultural productions, and the technical and aesthetic functioning of the train itself. In this way, the bullet train's spatial transformation of the Tōkaidō region resulted from the accumulation of arguments and agreements. Numerous interconnected contests

between JNR and various local interests, of which Kyoto was only one example, brought the bullet train into plans for restructuring the region and
its cities by redirecting mobility. But even before it opened for business,
planners' visions were already being challenged, as the system's technical
and aesthetic functions reshaped the landscape, and people began to view
the space in terms of their own evolving uses, desires, expectations, and
fears. Once built, the imagined space of the line was reshaped by the collective actions and outlooks of riders, some of whom wrote about their experiences, and urban planners, who incorporated high-speed rail into their visions of development for the region and beyond. Public musings about the
new accessibility between Tokyo and Osaka helped change contemporary
senses of distance and difference, transforming a national space composed
of culturally distinct regions into one that was becoming steadily smaller
and increasingly homogenized, focused on densely packed urban nodes
that were characterized and linked by high-tech infrastructures (a vision
that will be considered more closely in Chapter 3). As the bullet train was
taken up as a theme in popular culture, an alternative imagined space came
to overlay these two, attaching meanings to the space of the line connected
to tensions in contemporary Japanese society.[2] Space was a battleground, a
venue for the competition (not always conscious) among various sections of
state and society.

And, of course, the presence of the train itself, speeding across the landscape, whisking people from city to city, made a difference in how people
used and viewed the spaces it ran through, connected, damaged, and divided, reshaping mobility by facilitating some patterns of movement and
discouraging others.[3] The train reshaped space by creating new travel patterns and new approaches to cities, which people then considered, desired,
experienced, and discussed. In addition, its futuristic, stylish aesthetic became both a goal to strive for and a foil for counternarratives about contemporary Japanese society.[4] New meanings emerged around the bullet
train through the constant interaction of multiple factors, including ter-
rains along the route, sleekly designed cars equipped with technologies that
provided a safe, smooth, and quiet high-speed ride, and admirers and critics shaping public opinion. The elements by which the bullet train shaped
social space are not fixed, but rather shift in an ongoing process of mutual influence. Solidifying trackbed and developing technologies enabled

higher speeds; patterns of use changed configurations of schedules and station stops; updated designs, new lines, and faster alternatives using more exciting technologies affected popular fascination.[5] Considering the space of the line circa 1964 provides a momentary glimpse of an infrastructure in a never-ending process of transformation, its meanings constantly negotiated through a process of mutual construction among multiple agents.

RESHAPING CITIES: TOKYO AND OSAKA

Planning for the bullet train took place in the context of a long trend toward urbanization, especially the concentration of industry and economic activity in the Tokyo area. The resulting strain on the capital of rising population density (and the distress felt elsewhere due to population decline and uneven economic growth) inspired proposals for the reshaping of Japan and its cities, with the aim of redirecting flows of people and goods. The bullet train, along with other advances in transportation technology, was central to the discussion. Scholars involved in urban planning contributed their own visions of how high-speed rail could change urban and regional spaces. As the two largest cities and endpoints of the route, Tokyo and Osaka received the greatest attention. In Tokyo, primary concerns had to do with reducing or managing overpopulation. In Osaka, there were hopes that the train would raise the profile of Japan's "second city" by spurring local industry and decentralizing national government. But the train did not always cooperate with planners, interacting in unpredictable ways with people and other infrastructures, such as the new expressways planned and built at the same time through roughly the same places and the monorail between Haneda Airport and central Tokyo, along with extensive construction for the 1964 Tokyo Olympics. As an important factor in urban space and regional mobility, the bullet train impacted conceptions of space, changing the forms and identities of Tōkaidō cities.

In Tokyo—the seat of national government, economic center, and hub of the railroad system—the bullet train fostered the development of a layered urban space that could incorporate new transportation infrastructure to facilitate mobility within and beyond the city, including elevated tracks and roads above the existing cityscape, as well as underground passageways and shopping areas below it. In Yokohama and Osaka, the line had the opposite effect, stretching those cities out to encompass new stations on the outskirts. The physical changes wrought by the construction of the bullet train happened in and around the permanent manifestations of the line: its tracks and stations. And these were the focus of most attention for planners, including JNR employees, city officials, and academics.

As soon as the government approved the new line in December 1958, Tokyo city planners began to consider various possibilities for the location of the train's eastern terminal. In terms of railroad operations, Tokyo Station seemed to present the best option, as passengers could conveniently transfer to local trains and the expanding subway system at nearby Ōtemachi, which had served since the early twentieth century as a symbolic gateway into the capital.[6] But in terms of avoiding disruption in the dense heart of the city, less central sites made more sense. For instance, Shimbashi Station, two stops from Tokyo on the Yamanote loop line, was the terminal of Japan's first railroad line, opened in 1872 between Tokyo and Yokohama, and thus had historical precedent as the terminal of a railroad heading out of town along the Tōkaidō route. Other stations on the southwestern side of the Yamanote Line—Shinagawa, Shibuya, Shinjuku—were recommended for their geographical position and the convenience of transferring to local lines, so that arriving passengers could most easily get to their ultimate destinations within the city.[7] Access to the city's local mass transit systems was especially important in terms of the competition between the railroad and the growing airline industry. With the new monorail connecting the city center to Haneda Airport opening in the same year as the bullet train, access to the capital via air was about to become faster and more convenient, putting pressure on JNR to compete.

The main factor working against the choice of Tokyo Station as the terminal was the issue of land acquisition. While the difficulty of acquiring land for tracks and stations plagued the project all along the line, the challenge was greatest within Tokyo, where land was the most scarce and expensive, with the highest concentration of existing homes and businesses that would be affected by its construction and operation. In the end, passengers' convenience won the day, and JNR decided to bring the line all the way to Tokyo Station. But the challenge of building through crowded neighborhoods contributed to an important physical change to the capital, in the area of Tokyo Station and all along the line within the metropolitan area: the creation of a layered city. Planners in several corners—from JNR's head of local construction to architects and scholars of urban planning—began to see architectural layers as the solution to many of Tokyo's problems, beyond just the bullet train tracks.

Arguments for more innovative use of Tokyo's three-dimensional space, especially in connection with transportation infrastructures, began to proliferate as the line was being designed. In January 1959, the month after the government approved the new line, the *Yomiuri* newspaper sponsored a discussion among urban planners as part of a series titled "Tokyo in 10 Years." Takayama Eika, a professor of urban planning at Tokyo University and a leading figure in the field, pointed to the location of the bullet train terminal as the most significant issue facing the city and advocated a layering approach to development, stating, "Technologically, it would be efficient . . . to put in and improve facilities that use three dimensions in a more confined area. In other words, roads and parks take up a lot of space, so build them vertically. That is my best idea for the reconstruction of Tokyo."[8] Because of legal restrictions, earthquake fears, and the nascent state of modern earthquake engineering, Tokyo was still largely a low-rise city at the time. The Hotel New Otani would become the city's tallest building at seventeen floors when construction was completed in 1964, and Japan's first real high-rise office building was built four years later, the start of a boom in high-rise construction.[9] Faced with Tokyo's expanding sprawl, planners looked to layers and vertical construction as a solution.

Acclaimed architect Tange Kenzō was another early proponent of layered urban development, notably in the unusual and ambitious "Plan for Tokyo, 1960."[10] The team that created this plan included founding members of the Metabolist movement, whose concept of designing for adaptability to change and "the overlay of large urban projects onto existing dense fabrics" had a strong influence on future development strategies for Tokyo and held particular relevance for the construction of the bullet train through the dense fabric of the area surrounding Tokyo Station.[11] The plan was a response to a demographic crisis: the population of Tokyo was approaching ten million. A city of this size, Tange explains, cannot survive continuation of the existing centripetal pattern of development—in which the city expands outward from the center. Therefore, he calls for a shift to a linear pattern, in which functional nodes of commercial, residential, and other spaces extend in a line from the city center across Tokyo Bay. These clusters

be connected by a three-level system of transportation, divided ac-
g to speed of traffic and connected by a series of ramps. The bullet
iad only recently been approved and was still in the early planning
stages when Tange's group created this design, but they propose connecting their multilevel transportation system to the Tōkaidō and other lines
via subway, bringing yet another layer into this complex system. The creation of artificial ground was one of the Metabolists' responses to building
in a crowded urban area, and Tange's plan envisions the "ground" as a two-
level space: a public space for movement below and a quiet space for living
and working above. Citing buildings he had designed to create such layers,
he concludes, "This sort of solution appears to be one of the most promising
means of redeveloping urban areas."[12]

The complicated models and graphics that accompanied the published
versions of the plan—especially, perhaps, the idea of building the city out
over Tokyo Bay—hint at the impracticality of Tange's vision. In fact, the introduction to the plan in *Japan Architect*, the international edition of the
Japanese architectural journal *Shinkenchiku* (*New Architecture*), notes that
urban planning officials generally dismissed it as "visionary, but impracti-
cal." However, this comment is followed by a call for "visionary" and "practical" planners to communicate and incorporate each other's best qualities.[13]
And the emerging plans for the bullet train suggested that, to some extent,
the Tokyo of the future would require the sort of layers that Takayama and
Tange's team advocated.

The bullet train fit into and used this vision of a layered city and, once
built, illustrated its effectiveness. In some sense, it even helped compel it:
in Tange's view, the need for change was being driven by such advances
in transportation technology. "Mobility," he argued, "determines the structure of the city."[14] The presence of ten million people in the capital, and
their ability to speed to and from Nagoya, Kyoto, and Osaka not only in
trains but also in a growing number of cars on an expanding system of expressways, was the determining factor for Tange and other urban planners
in envisioning the city of the future. In later writings (further examined in
Chapter 3), Tange posited the Tōkaidō megalopolis as a model for urban
development on a national scale and called for "compact" urban designs
that could be achieved only through a rejection of what he identified as a
human tendency to cling to the ground. This plan, too, was fundamentally

about layers, calling for the use of elevators as vertical elements of the public road system.[15]

Though Tange's specific vision was never implemented, the bullet train propelled the layering of central Tokyo through the construction of elevated tracks and roads above the existing cityscape and underground passageways and shopping areas below it. With construction of the new line, Tokyo Station's shopping area, already a point of congestion, was moved underground in order to distribute the rising flow of foot traffic. People would enter the station at ground level, then descend to shops and subways below or ascend to elevated tracks for trains around and out of the city. A corridor of "name-brand shops" was moved below ground, then steadily expanded to create a subterranean city below the station.[16] In the center of Tokyo, bullet train tracks were elevated so as to minimize displacement. The head of bullet train construction in Tokyo gave the example of "Sushi Shop Alley" in Yūrakuchō as a neighborhood crowded with businesses that put tight constraints on both the path and construction process.[17] That crowded cluster of tiny sushi shops was swept away the following year with the construction of the Tokyo Transportation Hall, a commercial building topped by a revolving restaurant, but in Yūrakuchō and similarly crowded areas, the bullet train continues to this day to speed over long strings of small restaurants and bars tucked into the space beneath the tracks. These spaces are part of the "constructed underground" that Suzanne Mooney finds under the elevated tracks all around the Yamanote Line: "As the layers of the fabric of the city multiply ever upwards, the streets themselves are destined to take on the appearance and atmosphere of underground spaces."[18] Jeffrey Hou argues that such "mundane structures and shops embody the vision and reality of urban metabolism," as they are constantly adapting to their changing surroundings.[19] The bullet train, contributing to the layering of Tokyo, had a similar impact on neighborhoods like Yūrakuchō, where its tracks spurred new uses of space and created novel relationships between local populations and transportation infrastructure.

A November 1963 photo essay tracing the progress of bullet train construction highlighted this layering effect and categorized it as one of the singular achievements surrounding bullet train construction. Explanations accompanying aerial photographs taken all along the line noted that the bullet train would run on elevated tracks over a freight line from Shinagawa

Train as a symbol

to the southern corner of Tokyo. When this doubled line intersected the
Ōimachi Line, therefore, it would be a three-layer crossing. Further along,
it would cross the expressway to Osaka, at which point there was a subway
running underneath. This, the author proclaimed, would create "the world's
only four-layer crossing."[20] Such crowing over the esoteric (and somewhat
suspect) distinction of having the world's most complex railroad crossing is
just one of the sillier examples of the ubiquitous rhetoric of global superlatives in discussions of the new train—"world's first," "world's only," "world's
fastest"—that sought to place Japan on the cutting edge not only of a globally shared technology, but also of the field of urban planning. Thus, the idea
of layers not only informed the space conceived by Tokyo's urban planners
but also helped with the contemporary reinvention of Japan's international
identity.

Urban planners at the other end of the line had different concerns, related to their own city's particular problems. For Osaka, the most important issue had to do with the relative positions of the two terminal cities in
the national economy and in an urban hierarchy that placed Tokyo at the
apex and relegated Osaka to a position well below it in the context of ongoing centralization. Osaka's secondary status was a factor of not only population statistics but also economic productivity and growth rate, cultural
role, and popular perceptions. When discussion of the bullet train turned
to its impact on the cities of the Tōkaidō, the main question was often how
it would affect the flows of people, capital, goods, and ideas that helped determine a city's economic strength and cultural cachet. Would the bullet
train reduce or reinforce the centrality of Tokyo in national geographies
of population, governance, economics, and culture? Planners in the Osaka
area envisioned a leveled national and regional space, in which Osaka re-
gained some of its lost urban identity and status, and hence attracted both
people and economic activity.

From the point of view of Tokyo, this was not necessarily a matter of
competition. In many ways, the planners in Tokyo and Osaka had com-
plementary goals. In fact, urban planning experts in the capital hoped the
bullet train would encourage the outward flow of people. At the time the
bullet train was being planned, scholars and urban planners disagreed
on whether added mobility would encourage population movement toward or away from Tokyo, that is, whether it would mitigate or intensify

centralizing tendencies. In the 1959 *Yomiuri*-sponsored discussion of "Tokyo in 10 Years," participants held up the growing population as an issue of major concern. One theory about the impact of the bullet train on population was that increased speed and capacity for travel to points westward would encourage people to move out of Tokyo. But Isomura Eiichi, a professor of urban sociology at Tokyo Metropolitan University, suggested that improvements in transportation infrastructure—the bullet train, as well as new expressways and subways in various stages of planning and construction—would not, as expected, push people out to the suburbs but rather bring more in to the city. "It is unreasonable," he argued, "to make it convenient and then say, 'Get out!'"[21] As a visiting scholar at Harvard in 1957, Isomura had been greatly impressed by Jean Gottmann's notion of the Eastern Seaboard of the United States as a megalopolis, and he soon began to think of the Tōkaidō region in a similar way. His view of the bullet train's potential impact was shaped by the idea that the proliferation of transportation and communications infrastructures across this densely populated area would contribute to its unification almost into a single, massive city.[22]

The impact of high-speed rail on urban populations is related to the nature of railways, which tend to foster centralization around station areas. In the case of the New Tōkaidō Line, as a single line segment with endpoints in two major cities, it had the potential only to pull some weight away from Tokyo and toward cities farther west. However, the structure of the bullet train network that later expanded out from this initial line tended to reinforce the centrality of Tokyo, which became its hub.[23] All lines to the west connected in a single path between Tokyo and Kagoshima at the southern end of Kyūshū, and all lines to the north merged to end at the Tokyo terminal. In 1964 and for half a century after, Tokyo remained the focus of the high-speed rail network, limiting any decentralizing potential.

Many factors beyond transportation infrastructure contribute to demographic trends, but it is worth noting that Tokyo's population continued to rise after the bullet train began operation, though the rate of growth continued a gradual decline that had begun around the end of the occupation in 1952. The population of Tokyo as a percentage of national population, which had been on the rise, held relatively steady at a mid-1960s peak of about 11 percent with a mild decline beginning in 1970.[24] The three big Tōkaidō cities of Tokyo, Osaka, and Nagoya all experienced the same migration pattern in the postwar decades: rising rates of net in-migration peaked in 1961, then fell fairly steadily, to a net loss for both Osaka and Nagoya by the mid-1970s (but continued growth in Tokyo).[25] Five years after the inaugural run, Kyoto University professor Amano Kōzō used econometric modeling to single out the specific impact of the bullet train itself and found that it had intensified urbanization throughout the entire region.[26] A few years later, economist Chikaraishi Sadakazu critiqued the national development policies of the 1960s—in which the formation of a single Tōkaidō megalopolis enveloping Japan's major cities came to be seen as the model for future development—as having led to the "monstrous" growth of the region and deterioration of smaller cities. He found the cause of this failure in the assumption that infrastructural development would contribute to population dispersal, when in fact it always had an ambiguous potential in relation to population movement and could just as easily lead (as perhaps it did) to greater concentration in Tokyo.[27]

Expectations about the train's impact on population growth shaped perceptions of its potential to affect the relative positions of Tokyo and Osaka in both the national economy and the popular imagination. In the early decades of Japan's modern industrialization, from the 1870s to the 1920s, the urban "division of labor" echoed that of the preceding Tokugawa period, in which governmental functions were concentrated in Tokyo (then called Edo), while Osaka served as the center of business interests and a Japan-wide economy. A shift of economic activity from Osaka to Tokyo that began slowly with nineteenth-century industrialization gained momentum from Tokyo-centered military industrialization in the 1930s. That trend intensified with postwar recovery and economic growth in the 1950s. So for people in the Osaka area, it was easy to see the bullet train as a means for the city to regain its former position or at least slow the rate of change.[28]

But hopes for a local economic boost were balanced by other concerns. Economist Hoshino Yoshirō, writing over a year after the new line opened, incorporated its initial effects into his analysis of the potential economic impact on Kansai (the area surrounding Osaka and Kyoto) of the bullet train, the Osaka Expo planned for 1970, and the Meishin Expressway, which had opened the previous year between Nagoya and Kobe. Like Isomura six years before, he challenged optimistic views that new infrastructure would combat centralization, stating that the expressway, when connected eventually with Tokyo, "will not only further stabilize Tokyo's already dominant position in national politics and economy, but also, rather than raising the relative economic position of Kansai, may have the effect of absorbing Kansai into Tokyo."[29] The bullet train, Hoshino warned, was demonstrating such an effect, with businesses already moving their headquarters to Tokyo. That dynamic exemplified the impact of the bullet train on the space of the Tōkaidō region: planners expected that the train would draw people out of Tokyo, but the complex mesh of infrastructural, economic, and personal factors brought mixed results. Hoshino concluded that, with transportation networks tying the regions of Japan together more closely, economic growth must be planned not for Kansai in isolation, but for the nation as a whole. A schism can be seen emerging here between planners who expected high-speed rail to balance the distribution of population and industry and those who foresaw greater centralization. This point of disagreement, which persisted into later debates over expansion of the bullet train system, shaped expectations about the impact that high-speed rail would have on the Tōkaidō region and its cities.[30]

Hoshino pairs the bullet train with the Meishin Expressway because these major infrastructural projects were being planned and built at the same time in response to the same daunting industrial and economic challenges. New expressways connecting Tokyo, Nagoya, and Osaka also contributed to the social construction of space in Japan, simultaneously with the bullet train and often in parallel and reinforcing ways. Some felt that in the dawning age of the automobile, the highway system would dampen the significance of the bullet train, but usually high-speed rail and roads were viewed together as two parts of a growing transportation system that was spurring spatial changes.[31] Their interaction helped shape the ways in which politicians and urban planners viewed the space of the Tōkaidō. Hoshino's argument, like Isomura's, reflects the sense that improvements in transportation infrastructure were eliminating regions as separate entities, so that effective planning must view the Tōkaidō as an integrated whole within the nation. Today, while there is a movement to shift some government agencies to other cities, Japan's economy, industry, and government are still heavily centered in Tokyo. But the bullet train helped people rethink Japan's urban network and became one factor in both efforts at regional promotion and the realization that, more than ever, the various parts of Japan were bound tightly together. Planners foresaw strong transportation infrastructures creating a more evenly developed and economically vibrant Tōkaidō region.

Hoshino's and Isomura's doubts about the line's decentralizing effects seem to have been justified, and Tange's vision of accommodating the rising speed and volume of mobility by redirecting the expansion of Tokyo to follow a lengthening line, rather than an expanding circle, was never realized. But the bullet train did change the shape of some of the cities it connected, in exactly the manner Tange had warned was unsustainable: by creating new urban subcenters. Stations located on the outskirts of Yokohama and Osaka exerted an attractive force, causing the cities to bulge outward to encompass these transportation hubs. For the Yokohama stop, a new station was built far outside the urban area, where the train cut through a wide stretch of paddy fields (fig. 1). A discussion among government and JNR officials held on a running test train soon before the line opened to the public gives a sense of the agricultural character of the area prior to the construction of the bullet train. Riding along the route, participants commented on operational details and various difficulties related to construction. As the train approached New Yokohama Station, an editorial comment set the scene: "Passing through a section of embankments through many green fields . . . we arrive at New Yokohama Station."[32] Suddenly, the previously talkative officials find themselves with little to say. Nisugi Iwao, head of JNR's Construction Department, introduces the surprisingly green space to an incredulous Onouchi Yukio, head of the Roads Department in the Ministry of Construction, and Nagoya mayor Sugito Kiyoshi, who compares the location to Hashima, the site of the only bullet train station built where there was not a significant city.

NISUGI: "This station in the middle of paddy fields is New Yokohama Station."

ONOUCHI: "There's nothing here, is there?"

NISUGI: "Yeah, there's nothing here."

SUGITO: "It's like Hashima, isn't it? Is this the station for Yokohama?"[33]

The incongruity of a major high-speed railway station in a space devoid of roads and buildings was remarkable to people used to thinking about stations as busy, bustling areas in city centers. But contrary to the repeated claim that there was "nothing here," the station was, in fact, placed in the

新横浜駅　横浜線の菊名と小机両駅の中間に建設中で　将来第三京浜国道とこの附近で接続し　横浜の第二の中心となる　交差している　横浜線で　現在の横浜駅とはこの線を通じて連絡する計画になっている

FIGURE 1. New Yokohama Station under construction, c. 1963. (*Bungei shunjū*, November 1963.)

midst of cultivated land: a wide expanse of paddy fields. That agricultural character would soon be changed by the force of the bullet train station. Yokohama, a port city centered on the water, expanded to encompass a new inland center around the bullet train station, and those green fields were replaced by businesses, roads, and apartment buildings. The discursive work of transforming productive fields into "nothing" but empty space for development laid the conceptual groundwork for that transformation.

A similar process occurred in Osaka, though the location of the new station was not quite as rural as in Yokohama, so the change was less stark. Even so, a visitor to the station construction site about two years before its completion evoked the desolate atmosphere of land that had been stripped of its former productive functions in preparation for construction: "Dried-up fields all around. Not a soul in this vast landscape, where earth and sand pile up here and there into tomblike mounds. It is hard to imagine that this is the construction site for New Osaka Station, which by the time of the Olympics will be transformed into a great urban center."[34] This description in a local magazine published by the city government focused on the newly barren landscape surrounding the site but also evoked the grand plans envisioned for the area. In this way, like the officials' comments on the supposed emptiness of the productive but doomed fields around New Yokohama Station, it contributed to planners' presentation of the bullet train as "developing" empty spaces, rhetorical work that obfuscated the destruction caused by construction.

JNR selected the New Osaka Station site primarily on the basis of three factors: land purchases made decades earlier for a planned trunk line station in that area, the location's convenience for the anticipated westward extension of the system along the San'yō Line, and the difficulty and expense of acquiring land in the city center. Together with the construction of the Meishin Expressway and Senri New Town—a planned community about 10 to 15 kilometers north of the Osaka city center, developed in response to the 1963 New Urban Housing Development Act—the location formed part of a trend toward northward development, and local officials worked to transform the space across the Yodo River on the northern edge of the city into a transportation hub and urban subcenter through rezoning and the construction of roads, highways, bridges, and a subway line extension.[35] The resulting expansion of local transportation infrastructure

[margin note:] Stations in more rural areas had negative impacts

helped push urban development in a new direction. In this way, planning for Osaka echoed interwar efforts to control urban sprawl by extending planning to the peripheral space between city and country, but it differed in its focus on transportation infrastructure, rather than housing.[36]

The city rushed to complete local infrastructural expansion in time to greet the first bullet train. Just a week before its arrival, they celebrated the grand opening of an extension of the north-south Midōsuji subway line and a combined rail and road bridge across the Yodo River, an abandoned prewar project that had been reanimated by the decision to locate the bullet train station north of the city center. The subway extension itself inspired great excitement, with crowds gathering to buy commemorative tickets and ride the new length of rail on its first day. Moving above ground before crossing the Yodo River, the train then ran as an elevated line in the median of the New Midōsuji freeway across the river and northward. The local *Osaka shinbun* hailed the bridge itself as a new "famous place" (*shin meisho*) for its innovation in combining road and rail and the remarkable length of its arch and girders, and called the area around the station Osaka's "new north," connected by this "subway pipeline."[37] In spite of this timely construction, some challenges to access remained. Just two weeks earlier, the same newspaper had complained that the new terminal was "an exceedingly strange station. There's no road to get there."[38] With the aboveground subway extension blocking some roads, "even when the station is visible right in front of your eyes, you cannot see how to get there."[39] But in a sign that economic activity would soon encompass the new station, police found that crime syndicates were beginning to expand their activities geographically in order to establish their territory in this emerging urban center.[40]

Official city planners and legitimate businesses would soon catch up to organized crime. Osaka-based geographer Kobayashi Hiroshi enumerates the vast changes to the region in the decade following the opening of the new station: there was a remarkable increase in the number of businesses, entertainment establishments, warehouses, hotels, apartment buildings, and parking lots, while the numbers of factories and single-family homes declined precipitously, and the once-abundant farming areas disappeared completely. However, Kobayashi finds that the area did not develop to the extent planned, limited by the continued presence of a switchyard that effectively formed a border around the station. In addition, the on maintained a southward focus, with no entrance on its northern , a point of neglect that limited the area's potential.[41]

These examples suggest both the bullet train's usefulness and its unwieldiness as a tool for urban planning. Its construction furthered concrete, localized plans when it formed part of an integrated set of actions, as in the layering of Tokyo and, to some extent, the northward expansion of Osaka. However, it could not achieve the grander and more difficult goal of decentralization that some had imagined. Planners may control space in profound ways, but their control is never complete. The complicated process of deciding the route between Tokyo and Osaka shows that the social construction of space depends also on the individual decisions, actions, and perceptions of those who inhabit and use it.

DRAWING THE LINE: PLANNERS, POLITICIANS, AND PROTESTERS

Planners' success in realizing the space conceived in blueprints for the bullet train was constrained by popular protests and pressure from local officials while it was being built, and it was mitigated again even after the train started operating, as popular representations assigned alternative, sometimes conflicting meanings to that space. A view of actual and fictional contestations of planners' visions shows the multiple forces at work in the social construction of the bullet train space. Juxtaposing real-life protests with their fictionalization in popular-culture materials shows how fact and fiction interacted to chip away at state goals. Officials and urban planners in the cities along the route viewed bullet train construction primarily in terms of addressing local concerns, resulting in a cascade of changes: local interests shifted the path of the train, which in turn reshaped the physical and conceptual spaces of the Tōkaidō cities. Kyoto's campaign for a bullet train station, detailed in the previous chapter, is the most prominent success story of local pressure affecting JNR decision making, but similar struggles took place all along the line. For instance, the site of Gifu-Hashima Station emerged from a complex battle, in which the movement to attract a bullet train station to the prefecture was fractured when JNR's preliminary plans ran counter to local urban planning initiatives, pitting neighboring towns against each other in their efforts to influence railway planners.

Pressure on JNR from Gifu Prefecture was effective because it had political force behind it and was aided by geographic and cultural factors. But
the cause was complicated by competing local interests. JNR's early plans
had the train running in a relatively straight line from Nagoya to Sekigahara with a possible station (if at all) in the Kuwabara district of Hashima,
about 15 to 20 kilometers south of Ōgaki and Gifu Stations on the Tōkaidō
Main Line. The announcement of this route caused consternation among
business circles in Gifu City, who predicted the "localization" of the current station, as travelers bypassed the city on the new bullet train. An opposition movement began, and a petition signed by over half of the city's
residents called on JNR to shift the line 7 kilometers to the north (a request that may have been related to the city's plan to expand southward).
At the same time, Ōgaki, a smaller city west of Gifu, pressed for a different northern plan, following the Meishin Expressway. But Hashima City,
which was bisected by the expressway, initially opposed a new station. This
meant that three Gifu Prefecture cities each pushed for a different route.
The cities eventually agreed to leave the decision to Ōno Banboku, a leader
of the ruling party who was born in the prefecture, and they settled on promoting the northernmost route (see inset in map 1, p. xii).[42]

In their efforts to convince JNR to adopt their "Gifu People's Plan,"
petitioners enumerated several factors favoring a northern route, none
of which had to do with the prefecture's own economic interests. Some
of these were related to the preservation of important historic, religious,
and agricultural sites (the site of the 1600 Battle of Sekigahara, the Nangū
Shrine, and wet paddy fields) that would be damaged by the JNR plan. Another set of reasons highlighted the technical aspects of construction and
operation, appealing directly to the railway agency's concerns about speed
and cost: JNR's route had a much steeper slope and would run through the
Yōrō and Ama districts, which had flooded during a 1959 typhoon, making
construction more expensive and operation more dangerous. In addition,
petitioners argued, a rail bridge across the Nagara River would limit the
added time and lower the costs of the northern route.[43]

Political pressures combined with geographical features (including
poor ground quality along JNR's proposed route, which cut across several
river deltas) to bring about a compromise between the two plans, with a
stop in Hashima. Ōno may have added his weight for this choice, leading

criticisms of Gifu-Hashima as a "political station."[44] The decision, announced in January 1961, and the approval of the route by the Ministry of Transportation, brought a new burst of petitions from Gifu, as well as the cities of Bisai and Ichinomiya in neighboring Aichi Prefecture.[45] JNR's insistence that no other appropriate route was possible left local opponents with few options other than to resist selling their land for the project. Newspapers predicted that land acquisition would, therefore, prove difficult.[46] Indeed, in these cities and other points along the line, resistance by individuals and local groups through refusal to sell land to JNR caused the construction budget to skyrocket and threatened to delay the opening.[47]

Relatively powerful entities like Kyoto and Gifu Prefectures were able to use their political weight to bring a significant change to JNR's plans, but elsewhere, popular opposition through small, fragmented acts of protest mostly failed to affect its path. The planned structures of the new line—both tracks and stations—were frequent sites of such contestation. In 1962, novelist and scriptwriter Shimada Kazuo traveled the length of the bullet train track under construction, hunting for plot ideas for the television drama *Bullet Train* (*Dangan ressha*), which was aired during prime evening viewing hours by Japan's national public broadcaster, Nippon Hōsō Kyōkai (NHK), just before the line opened. Visiting over fifty construction sites along the route, Shimada reported hearing the same thing everywhere: "Our technology will overcome mountains and rivers. What it cannot overcome is people. Building the bullet train is a battle with people."[48] Skirmishes involved not only protesters but also tensions with local communities and negotiations to buy necessary land. In a context in which land expropriation was considered to be an option of last resort, the power of families and individuals asserting their ownership rights and resisting government pressures created headaches for JNR. But they did not change the route.

Nagoya provides a good example of the quixotic battles sparked by bullet train planning. Popular enthusiasm for the project in Nagoya fostered a cooperative attitude toward JNR. Incorporating plans for the bullet train into an ongoing urban revitalization project in the area around the station, the city arranged JNR's purchase of the necessary land in that area.[49] But even in the midst of such popular and official welcome, opposition was still visible. On his journey along the line under construction, Shimada noticed

a signboard at the entrance to Nagoya Station that read, "Citizens ⸗
goya! Our town, which to us is the most affordable and enjoyable, i⸗
to be demolished by JNR. Together, let's staunchly protect our town."[50] The
enthusiasm for bringing high-speed rail to local places was not universal,
and what seemed beneficial to a city in the eyes of its political and economic leaders was not necessarily attractive to all of its residents. Opposition to the construction of the new line, in this case, was tied to a particular
view of the city and its future, one that would be damaged, rather than promoted, by having the bullet train stop there.

As Shimada's trip and the television shows it inspired demonstrate, in
addition to raising costs and hindering construction, these real-life com-
plications also provided grist for depictions of the new line in popular culture. Protests and petitions were not usually prominent enough to make
the national news, but they entered the public consciousness through fictionalized accounts presented to popular audiences through a media mix
of print, film, and television shows like Shimada's. Such popular narratives
can have a complicated effect on public perceptions. They are explicitly fictional, so even when they are based on actual events, the line between fact
and fiction is never obvious, and they cannot be (and most likely were not
at the time) taken as factual accounts. However, by putting events into the
form of emotional narratives, presented (in the cases of film and television) through good-looking actors and sometimes in color, popular-culture
treatments of these stories not only reached a broad national audience but
also elicited a gut reaction that is less likely to come from a brief article
in a local newspaper. Therefore, it is imperative to consider the impact of
such materials on popular views of the new line and its significance for the
nation.

While the viewing public got JNR's triumphant narrative from news reports and promotional materials, fiction writers and artists appropriated
the image of the bullet train to highlight central tensions in Japanese society, thereby creating a mental landscape within which people understood
and re-interpreted the space created by planners. Problems that plagued its
construction later shaped the space through their representation in cultural
materials that capitalized on the train's popularity by using it as a topic,
setting, or plot device. Writing about the bullet train could thus function
in support of protests or resistance against the new line and the spatial

The duality

transformations it seemed to be effecting. Running across the varied landscape of the Tōkaidō, the train created romantic images of modern technology in harmony with the natural and built environment, running smoothly
past fields and mountains, temple pagodas, and modern cityscapes. But the
frequent focus of popular-culture representations on corruption, greed,
and environmental or cultural damage undercut celebratory images of
smooth progress, proffering a darker vision of the processes of development symbolized by this new transportation infrastructure. In some ways,
this echoed early twentieth-century writings that decried the expansion of
the railroad system, depicting trains as monstrous and terrifying.[51] In contrast to those critiques, which presented dangers inherent in the train itself, negative images of the bullet train tended to focus on human elements,
exploring associated social, economic, and political inequalities. This difference may stem from the centrality of railroads to everyday life by the
1960s, in contrast to the earlier period, when trains were still relatively new
to many Japanese.[52] Writers criticizing the bullet train were fighting a battle not against the machine but against other people.

The ten episodes of the NHK television show *Bullet Train* that emerged
from Shimada's plot-hunting expedition capture the possibilities opened up
by the economic development that the bullet train represents but also the
potential pitfalls. Together, they depict several forms of popular resistance
against construction of the new line. Each of these conflicts is resolved by
the end of the episode, and the series is basically a celebration of the line as
a source of national pride. This effect is achieved in part through use of actual footage of construction sites and test runs, contributing to excitement
about the train and popular desire to ride it by showing it speeding across
the countryside, bursting out of a tunnel, or pulling into a brand-new station. But JNR does not come away unscathed. The depiction of harm to local communities and individuals resonated with the real-life stories that
cropped up in newspapers and magazines to fix in the public mind the image of damage to the places where the new line was passing through and
underscore the prioritizing of national over individual or community interests. For instance, one episode, titled "Station amid the Rice Paddies,"
is set in Hashima and directly confronts the difficulties and tensions surrounding land acquisition. The show does some work in JNR's favor. For
instance, in an exasperated response to a reporter's question about rumors

that the placement of Gifu-Hashima Station was a matter of pork barrel politics, one character lays out in great detail the environmental, technical, and operational reasons for the northward shift of the line that brought the station to this rural area. But the rest of the episode is more ambiguous.

Opening shots of farmhouses, temples, and shrines, accompanied by a voice-over describing Hashima as a small agricultural town, sets the scene. After hearing news of the new bullet train station, three young residents—Kiyoshi, his sister Yoshiko, and their friend Hiromu—talk about whether it will be good or bad for their town. While they are optimistic about better access to Tokyo and higher land prices, they also worry about damage to agricultural fields and their own future as farmers. Some scenes show enthusiastic planning for urban and economic growth, but most of the action centers on the contrasting attitudes, choices, and experiences of the young friends. Hiromu and his family, whose land lies in the vicinity of the planned station, resist the intense badgering of a broker trying to purchase it, but Kiyoshi, dazzled by the silver-tongued broker, goes against his mother's wishes and sells their property. He uses the proceeds to build a factory, but by the end of the episode, he has gone bankrupt and leaves Hashima in the hopes of finding work in Tokyo, asking Hiromu to care for his mother and sister in his absence. The episode is sprinkled with some of the tropes of JNR's triumphant narrative, including actual footage of construction, praise for the workers whose diligence made up for delays in choosing the site to finish in time for test runs, and the successful construction of the station, a "concrete castle that gradually rose up amid the rice paddies, . . . certain proof that a new era was arriving."[53] But the story as a whole carries a strong message about the pain, dislocation, and damage inflicted on communities, families, individuals, and their land. Echoing news stories about the difficulties of land acquisition, the episode would be more likely to generate sympathy for the farmers than for JNR or its brokers.

Another episode told a story of dishonest brokers taking advantage of bullet train construction to make fast money through land speculation. This was set against the background of general opposition to the new line in a rural town where the train would pass through without stopping. In this episode, an arriving JNR construction crew meets resentment and anger from residents of an unnamed town outside Kyoto, who had resisted selling their land but eventually gave in under pressure. The local people

complain that the lack of an express stop on the existing line is slowly killing their town and predict that the bullet train will only compound their
problems. Their antipathy toward the project leads them to blame the JNR
interlopers for a rash of petty crimes—fields have been damaged, and vege-
tables, chickens, and eggs are being pilfered from farms. But they soon discover that those crimes were in fact orchestrated by unscrupulous land brokers seeking to drive down prices, and the assistance of JNR employees in
catching the actual criminals brings a sudden change in attitudes toward
the railway agency and its workers.[54] In fact, a dishonest land broker actually had been arrested for a scam related to land purchases for the bullet
train in the city of Hikone in Shiga Prefecture, the vicinity of the episode's
setting.[55] Drawing on real cases of speculation and opposition that plagued
construction, stories like this helped build awareness of negative facets of
the bullet train, contrasting with JNR's airbrushed representations.

The fact that this episode's crooked broker planned to build a super-
market and housing to replace small shops and farmland highlights another set of real-life tensions surrounding this infrastructural project: the
importance of cultural and environmental preservation versus industrial
development and economic growth. While many landowners' refusal to
sell stemmed from either personal attachment to a place or greed, resistance in some cases was based on fear of cultural obliteration, especially
due to environmental and other harms in places where the train would race
through without stopping. Such places would miss out on the benefits of
having a bullet train station but would nevertheless have to deal with the
destruction of existing space for tracks and continuing nuisances associated with operation of the train, such as noise pollution and vibrations.

The show's writers illustrated this concern most vividly in an episode devoted to the topic of tunnels, a key element in achieving the relatively straight
line required for JNR's speed goals. An opening voice-over—accompanying
an image of the train running across a valley and into a tunnel, then out
the other side—lauds tunnels as "the road to rail modernization."[56] But the
narrative turns quickly to focus on a tunnel-boring project that elicits local animosity. In the story, the line's planned path leads through a mountain that, according to local legend, houses a pair of dragons who protect the
nearby *mikan* orange farms, keeping them healthy and productive, even in
times of drought. The legend derives from the presence of an underground

aquifer, which local leaders fear might be damaged by tunnel construction, ruining their farms and hence their livelihood and community. As groundwater begins to seep through the tunnel walls, the JNR crew struggles to balance unrelenting demands from the agency's leadership to maintain the construction schedule with mounting evidence of a serious problem. Inevitably, water comes pouring into the tunnel, and all the wells in the area go dry. The show manages to wring a happy ending out of this disaster by depicting the villagers cooperating with the railroad workers to bring in sandbags and waterproof the tunnel, but the narrative sets up JNR as callous and the bullet train as a potential threat to the community, its economic base, and the particular local culture of "Two Dragon Mountain."[57]

The *Bullet Train* drama aired on national television at a time when sales of television sets had exploded, and about two thirds of Japanese households owned one.[58] Newspaper articles promoted the show as exciting for its impressive footage of construction and test runs, and for shooting episodes on real JNR construction sites, making for a "completely new method of producing a documentary drama."[59] Though information about viewership is not available, the venue, broadcast time, and prominent promotion of the show suggest that many probably saw it. Therefore, the idealized images of the train's speed, elegant design, and victories of construction and technology presented by JNR and often repeated in news reports would have been balanced in the public eye with these tales of the public nuisances and private pain that it caused. And similar images were conveyed through other media, such as novels and films.[60]

Among the conflicts Shimada found on his plot-hunting trip for the show were cases of people fighting against the track cutting through rice paddies, abutting important buildings and landscapes, or just traversing residential areas where its noise and vibrations would impact the daily lives of people who lived there. He gave a specific example of the forest surrounding the Oiso Shrine in Shiga Prefecture, which was sliced in two by the tracks.[61] The forest was a powerful symbol for critiques of the privileging of urban over rural space, and its status as a "sacred space" underlined the tension between the national goals of development and economic growth on the one hand and the value of preserving nature and tradition on the other. The Oiso Forest functioned similarly as an emblem of reckless development in a different kind of cultural product. Travel writer Nagasawa

Kikuya detailed every aspect of riding the New Tōkaidō Line in his unusual *Shinkansen Travel Notes* (*Shinkansen ryokō memo*). Breaking from the travel guide as a genre, Nagasawa, a longtime critic of JNR, makes the train the villain of its own story, consistently highlighting its drawbacks and contrasting it negatively with the Tōkaidō Main Line. In its focus on the damage done by the new line to both the landscape and the experience of travel, Nagasawa's guide resonated with the *Bullet Train* television show.

Nagasawa first mentions the Oiso Forest to illustrate the important and interesting sights that riders are likely to miss because of the train's high speed, which reduces nearby elements of the landscape to just a momentary blur. Taking a quick swipe at "the coercive measures used to allow the Shinkansen to pass through the precincts of a shrine," he later returns to this example in order to consider at a more leisurely pace the conflict between development and preservation.[62] Nagasawa proclaims his scorn for those who, out of self-interest or convenience, opportunistically engage in activities resulting in the destruction of culture. But he acknowledges that open space for development was becoming increasingly hard to find, and wonders, indeed, whether this particular forest would have been able to maintain its value as a poetic place regardless of the bullet train's intrusion, given that conditions around it were rapidly changing.[63] Thus, his mourning of a specific injury becomes a more general critique of development without regard for the preservation of important historical and cultural spaces and a meditation on the conflict between economic growth and cultural preservation, an issue of growing importance for the Japanese people in the years of high-speed economic growth. The bullet train and the Oiso Forest worked at cross-purposes in shaping public imagination: as the running train created compelling images of advanced technology traversing bucolic settings, the desecrated forest reframed it as an enemy of Japanese tradition and nature.

Nagasawa's travel guide highlights another point of frustration with the bullet train, related to a change that Tim Ingold identifies as a characteristic of modernity: the transformation of the solid line formed by the act of travel into a fragmented set of points. Ingold describes a difference between the trails created by two types of travel, which he characterizes as wayfaring and transport. The wayfarer engages with the landscape while moving through it, forming a lengthening line. The transported passenger, in contrast, skims over the surface, traveling not through but rather across space, creating a connection between two points, with all activity concentrated on those points.[64] The railway as a form of transport certainly played a role in this change, forcing travelers into predetermined patterns of movement, from specific starting points to specific destinations, while removing them from contact with the spaces in between. In his study of Europe's earliest trains, Wolfgang Schivelbusch argues that the first railways, therefore, changed the traveler's relation to the landscape being traversed, taking away the opportunity to "savor" the spaces in between stations.[65]

Rail travel enthusiasts had partially repaired that rupture between traveler and landscape with the development of "train window" tourism. But the bullet train intensified passengers' alienation from the spaces traversed by creating a pattern with far fewer points separated by much greater distances and by further separating the traveler from the intervening landscape, which was blurred by the train's higher speeds. Nagasawa mourned that renewed loss of connection to the land, repeatedly complaining that the bullet train's speed prevented the rider from spotting landmarks that were familiar from rides on the existing Tōkaidō Line. On the slower train, passengers could not physically engage with the land along the tracks, but at least they could participate in a limited sort of tourism by viewing and enjoying points of interest along the way, which were carefully detailed in guidebooks.

Nagasawa contrasted this with the faster trains, writing somewhat ruefully, "From inside the car of a Hikari or Kodama, which pass by faster than you can say 'an instant,' one cannot gaze out the window in the leisurely way of the past."[66] This he halfheartedly justifies by pointing out that these two trains—the super-express and limited express, respectively—were intended primarily for business use and, therefore, aimed to carry passengers to their destinations quickly, rather than allowing them to view the scenery. Even so, he grumbled about all the familiar sights he missed seeing. For instance, "The rail bridge over the Seta River flowing out of Lake Biwa is an enjoyable view for rail passengers, and as the herald that one is about to come to Ōtsu and Kyoto, it has a very nostalgic feel, but now, unless you get a lot of warning ahead of time, you just pass right by in a flash, and by the time you notice, it is already behind you."[67] For Nagasawa, the balance of the bullet train's impact was loss, a disappearance of valuable spaces and

ɪjoyable experiences due to the shrinkage effected by high-speed transportation. Such popular perspectives on the bullet train show that negative
views of the new line were part of the cultural imaginary, helping to shape
the social space around the line, if not its actual path.

JNR's celebratory view and its rejection are juxtaposed to powerful effect in another popular fictional story about the corruption scandals that
tainted the project: Kajiyama Toshiyuki's 1963 detective novel *Dream
Super-Express* (*Yume no chōtokkyū*) and a film adaptation made the following year. These works will be examined closely in the next chapter but are
briefly mentioned here because the runaway success of both novel and film
suggests that counternarratives to the official story of technological progress and economic development through fast and efficient transportation
were very much part of the popular understanding. Together, the novel
and film helped spread a multimedia message that would shape the public imagination of the spaces of the new line. This best-selling tale of corruption in bullet train planning depicted the line as a space of intrigue and
crime. Based on a news story about the arrest of several people for some
shady land deals surrounding bullet train construction, they portray the
dark underbelly of the bright, modern society seen in celebrations of the
line.[68] In the film's final scene, the antihero is riding the old Tōkaidō Line
and reading a newspaper report of the death of his co-conspirator, killed in
the course of their quest for money and revenge. We glimpse the headline:
"Sudden Turn in Condominium Murder Incident: Developing into Shinkansen Corporation Graft." When the passenger next to him comments
with awe and admiration on the bullet train visible in the distance, "Fast,
stylish . . . isn't the super-express great?" he responds angrily, slamming
down the window shade so he does not have to see the train, a symbol not
of progress, but of greed, deception, and murder.[69]

These examples of bullet train fiction depict the futility of resistance
against the institutional force of JNR. But their popularity itself amplified

opposition in a way that relatively powerless protesters could not. Though
protests rarely changed the course of the line, fictionalized stories of resistance contributed to the social construction of its space by affecting popular understanding of its significance to Japanese society. As Henri Lefebvre
argues, although the organization of space is a means of control, "it escapes
from those who would make use of it. The social and political (state) forces

which engendered this space now seek, but fail, to master it completely."[70]
Representations of the bullet train in popular culture helped to create and
disseminate alternatives to JNR's heroic narrative.

REIMAGINING THE TŌKAIDŌ:
NEW TRAVELERS, NEW TRIPS

The plans that emerged through the process of negotiation and compromise among JNR, local governments, and individual landowners became
the physical construction that transformed the landscape in tangible and
visible ways, and with these physical changes came a reimagining of those
spaces, not only through popular culture but also through the actual (and
potential) use of the train by millions of people. As an aesthetic object, the
bullet train excited the desire to experience its speed and elegance. As a new
framework for human mobility, it changed notions of proximity, time, and
place; it helped shape actions by creating possible "time-space routines."[71]
The public discussion surrounding its construction and opening reveals
new views of the regional space, the converging cultures of the Tōkaidō
creating a single megalopolis, while Tokyo and Osaka were becoming more
familiar to each other as close neighbors, just a quick train ride apart.

The bullet train changed not only the way that people saw other parts of
the country but also what they were likely to see. The fast trips and accompanying "speed-up" of the entire national rail system—through schedule
changes that timed expresses on connecting lines to meet the bullet train—
facilitated shorter trips to more distant destinations, inspiring new itineraries for day trips, weekend jaunts, school excursions, and honeymoons.
The perceived space of the Tōkaidō (the agglomeration of individual uses
and views of the train) was shaped in part by the itineraries and pack-
ages sold through travel agencies, which stoked the desires kindled by the
train's unprecedented speed and sleek design. Promoters of leisure travel
by rail at Japan Travel Bureau (JTB) depicted the new line as an inspiration for a journey.[72] The organization's monthly magazine, *Journey* (*Tabi*),
opened a November 1964 article on new itineraries using the bullet train
with an imagined quotation: "Just once, I want to ride the super-express—
there are now voices murmuring such words all over the country."[73] JTB
then fed that yearning, pointing out, "For tourist sites between Tokyo and
Osaka, . . . it has become much faster to get to places that can be reached by

alighting at ten new stations," and giving the example of the trip from To-kyo to the resort town of Atami, which was cut nearly in half to fifty-eight minutes.[74] The article details several "fresh speed plans" facilitated by the bullet train's compressed schedule, using connections to other lines. For in-stance, it had become possible to visit Himeji—the home of "Japan's most famous castle"—in just a day trip from Tokyo. JTB outlined a tight sched-ule that allowed for only three hours to enjoy the castle before catching the train for the first leg of the return journey, but the trip had become possible even for a busy Tokyoite with only one free day available.[75]

A more typical journey was between the terminal cities. JTB took ad-vantage of the excitement surrounding the opening of the new line to pro-mote a series of group trips centered on using the bullet train and other new transportation infrastructure. The introduction in *Journey* of a set of itineraries grouped under the heading "Trips to Commemorate the Open-ing of the Shinkansen" mixed stereotypical views of each city with the al-lure of new modes of transportation. "On a day trip using the super-express connecting Tokyo and New Osaka, [see] the urban beauty of modern To-kyo, which was remarkably transformed by hosting the Olympics, and visit the sights of the monorail, expressway, and Olympic facilities; or con-versely, Kyoto and Osaka have Kyoto's atmosphere of the circuit of famous gardens . . . and the new essential, a pleasant drive on the Meishin Express-way."[76] The goal of these trips was not just to view contemporary architec-ture or ancient gardens, but specifically to experience the speed and excite-ment of new modes of transportation: of course, the bullet train itself, but also the monorail connecting Haneda Airport to downtown Tokyo (sim-ilarly completed just in time for the Olympics) and the new expressways built to facilitate high-speed movement of the growing number of auto-mobiles on the roads of the Tōkaidō region. As a *Kyoto shinbun* editorial argued, improvements in transportation infrastructure transformed and sometimes impeded familiar sights, but they could also create new iconic views. For instance, the construction of the Ōhashi Bridge crossing Lake Biwa might have ruined the famous sight of "wild geese at Katada" (one of the "Eight Views of Omi" depicted in a series of nineteenth-century wood-block prints) but a new set of famous views of Lake Biwa had already been enumerated, and the editorial called for inclusion of the bridge itself on that list.[77] No longer just a way to get to a destination, road and rail themselves formed a part of the spectacle to be enjoyed by sightseers in the modern Japanese city.

The bullet train's technical function of shrinking time-distances and the resulting revision of how far a person could roam on a quick trip changed not only travel plans but also attitudes about other parts of Japan. The ability to get from one end of the Tōkaidō to the other and back within a day seems to have made people in each city feel closer together and more familiar, even if they themselves were not regularly speeding back and forth. Tokyo and Osaka came to be seen not as distant cities—the antipodes of the Tōkaidō—but rather as close neighbors, even two parts of a single megalopolis. This was not a novel dynamic. Considering the development of railroads in nineteenth-century Europe, for instance, Schivelbusch argues that the railroad changed perceptions of time and space, as faster transportation effectively shortened distances and expanded accessibility, making it easier for people in cities to visit remote places and for the products of those places to be brought to urban markets.[78] Steven Ericson notes a similar compression in people's sense of space accompanying late nineteenth-century railroad construction in Japan: the first Tōkaidō line, completed in 1889, reduced travel time between Tokyo and Kobe from a week by horse-drawn omnibus or a few days by steamship to just twenty hours by train.[79] And the development of railways and other transportation infrastructures in the early twentieth century promoted a sense of the expanding empire as a unified (if variegated) national space.[80]

The 1964 bullet train made a relatively small change, cutting the journey between Tokyo and Osaka, which had only recently been reduced from nine to under seven hours, down to three hours on the fastest Hikari superexpress. With the opening of the New Tōkaidō Line, a Tokyo businessman could take a day trip for a meeting in Osaka; a Kansai housewife could go shopping in Tokyo and be back in time to greet her husband when he came home from work; a foreign tourist could while away the morning in Kyoto's ancient temples, enjoy an afternoon sightseeing through the bullet train window, and admire the sunset from the revolving Sky Lounge of Tokyo's Hotel New Otani. While an extra few hours might seem relatively insignificant, fueled by the promotional machinery of JTB, the shrinking travel time and expanded capacity changed the way that people interacted with and thought about the Tōkaidō cities and the distance between them.

Pundits, celebrities, and other early bullet train riders mused publicly about these changes in the pages of newspapers and journals. A prominent factor was volume: with the convenience created by high-speed travel, more people could travel more frequently between the cities on the New Tōkaidō Line, thus tightening the bonds among them. Even as construction was being completed, there persisted among Osakans a sense that they were misunderstood by Tokyoites and the rest of Japan, but new patterns of travel and increased transportation capacity had the potential to create a sense of connectedness among the residents of cities along the line.[81] Some predicted that Tokyo would become a familiar place to more frequent visitors from Osaka, and interactions with the people of Osaka would be commonplace for residents of Tokyo. They imagined that the bullet train would create new configurations of the world by bringing into contact people who would not otherwise meet, or just make people in distant cities think of each other in a more familiar way.

Such visions of familiarity were budding even before the train was built. The *Yomiuri* newspaper predicted as much in its January 1959 "Tokyo in 10 Years" article series. One of these articles, considering the potential impact of the planned New Tōkaidō Line, imagined fictional residents of a Tokyo suburb, Mr. Shimagi and his wife, Ryōko, on their way home from Tokyo Station after seeing off their niece, who had embarked on the bullet train for her honeymoon. "As Shimagi and Ryōko left the station, a winter rain dampened their shoulders. Ryōko said, 'I wonder if it's raining in Osaka.' It's a fact that since the completion of the New Tōkaidō Line, Osaka has become part of the daily life of Tokyoites. In Shimagi's company, people from the Osaka branch, whom they rarely would have seen in the old days, come [to Tokyo] just about every day, and even in the household, the city of Osaka is frequently a topic of conversation. Flying down the rain-covered expressway, Shimagi promised to go as a family on a fun trip to Osaka the following Sunday."[82] Peering into the near future, the newspaper imagined that the fast train, whose construction had in reality only just been approved and not yet begun, and the expressway, which would take Tokyoites to the station from their suburban homes, would make Osaka a part of daily life for the average resident of Tokyo.

The comments of novelist and Osaka native Tanabe Seiko about a trip she took to Izu using the bullet train reflect a dwindling sense of distance and difference from the point of view of Kansai. Writing just after the line opened, she captured the ways in which it was changing her perception of the opposite end of the Tōkaidō. She wrote, "I don't know much about the eastern provinces. Within the Kansai sensibility, ours is the land of the imperial palace, and even one step beyond the area of the ancient capital are places that we think of as barbarian provinces. We stubbornly maintain images that float up from old names like Tōtōmi, Suruga, Kai, Izu, Sagami, all of which we think of as distant lands. But with the appearance of the bullet train, from a paper journey via the timetable, suddenly eastern Japan has ceased to be a distant land."[83] By referring to Kansai's history as the home of the imperial court until the Meiji emperor's move to the present capital and using archaic names for places between Osaka and Tokyo, Tanabe suggests that the distance of those places is also a thing of the past. And one need not even board the train to feel the closeness of previously distant lands, but simply take a "paper journey via the timetable." Just knowing how quickly one could get there was sufficient to make them feel close.

This sense of closeness generated a vision of the region's cities as a single unified space: a Tōkaidō megalopolis. Discussions of a potential regional administrative framework began at the start of the 1960s, while the line was still under construction.[84] Best-selling author Kajiyama Toshiyuki (by this time well known for the aforementioned detective novel about corruption in bullet train land acquisition), gives an anecdotal account of this idea's popularity in an article considering the new line's impact on Osaka. He recounts a conversation he had on the super-express with a businessman who advocated merging Tokyo and Osaka to make a single megacity. The businessman's argument was that because Tokyo was already the political center and was fast becoming the economic center, if trends continued, the city would become unlivable, with land prices soaring out of reach for most and traffic problems intensifying beyond practicability. Creating a huge "Tōkaidō City" encompassing Tokyo, Yokohama, Nagoya, Kyoto, Osaka, and Kobe, he claimed, would allow city planners to control urban development by assigning a specific role to each city: "Tokyo is the city of politics and government. Osaka is a merchant town, the Nagoya area is factory land, and Yokohama and Kobe are international ports."[85] With cities planned according to such an urban division of labor, the anonymous businessman claimed, "taxes will be lower, and everyone will live more comfortably."[86] Kajiyama describes himself as being fascinated by the huge scope of the idea and swept up by the man's logic, but ultimately skeptical that it could ever come to pass. He was taken aback when he introduced the idea at a bar one evening, and the majority of customers seemed to think such a thing might happen in the near future. "Think about it," they said. "Now, on the super-express, it's three and a half hours. On a plane, it's one hour. When the Tokyo-Nagoya expressway is completed, . . . Osaka will be closer to Tokyo."[87] With transportation facilities having drawn distant cities closer together, Kajiyama suggests, people were beginning not only to dream of them as a single unit but to accept that prospect as a fairly certain near-future reality.

Kajiyama raised the idea of a Tōkaidō megalopolis not to promote or endorse it but rather to express his skepticism about both its practicability and its merit. While shrinking distances opened up new opportunities, some effects were less desirable. Schivelbusch argues that in nineteenth-century Europe, while the railway's minimization of travel time and concomitant expansion of spatial accessibility clearly had some positive economic results (for instance, by bringing the products of remote places to metropolitan markets more quickly and cheaply), it also had a negative cultural impact on those previously inaccessible localities, which lost the particular value they had held as remote places—what Walter Benjamin encapsulates as "aura."[88] In the case of the Tōkaidō, local cultures and regional differences could potentially be highlighted by fast connections, but they were also threatened by accompanying Tokyo-centric homogenization.

Though Kajiyama had no personal connection to Osaka—he was raised in Japanese-occupied Seoul, repatriated to his father's hometown of Hiroshima after the war, and later moved to Tokyo—he fell in love with the city, even to the point of using the local dialect, a literary defense against homogenization on the city's behalf.[89] He praised its particular charms: the bustling but unpretentious crowds, the warm informality and friendly openness of its people, the devoted shopkeepers, the welcoming bars, and especially the food. But it bothered him that, with the influx of Tokyo culture (not just via trains but also television), "Osaka has gradually begun to lose the scent of old times."[90] Kajiyama bemoaned homogenizing trends in manners, customs, and language. Osaka conversations, he

wrote, were a jumble of Tokyo and Osaka dialects, while there was
ion for using Osaka dialect in Tokyo. Even as he expressed apprecia
the convenience that made Osaka "like the town next door" to Tc
also warned, "I do not want Osaka to become Tokyo-ized. I want Osaka
to always be Osaka. Tokyo is like a colony, so it's all right if it doesn't have
its own Tokyo scent. But Osaka, same as Kyoto, should have its own local scent. . . . I would like to have some essence of Osaka left remaining.
If not, I don't understand why one would go to Osaka. . . . I would like for
Osaka to have things Tokyo does not have. . . . When I get off the bullet
train and take the subway to Umeda, I want [to find] a city that is liberally
sprinkled with the scent of Osaka."[91] Kajiyama does not elaborate on what
makes Tokyo "like a colony," but the phrase suggests inundation by a foreign (in the case of postwar Tokyo, American) culture. According to this
logic, if Tokyo was allowed to be the cultural metropole of the nation, homogenizing outlying areas with its imposed foreign culture, then the rich
flavors of places like Osaka and the last traces of an authentic local Japan
would be lost. Kajiyama's views of colonization and cultural loss may be
colored by his own experience growing up in Japanese-occupied Seoul during a time when official Japanese policy was the complete assimilation of
Koreans within the Japanese empire through the eradication of Korean culture, a topic he would later grapple with in his writing. His mixed feelings
about the high-speed rail traversing the Tōkaidō signal the variegated response among the Japanese public. While the story of the first bullet train
is often told in terms of triumphant progress, for many Japanese, there were
drawbacks and downsides that came with, and sometimes overwhelmed,
the benefits.

The possibility that greater access to Tokyo would cause other cities to
lose their local flavor was a matter of concern in part because it seemed
already to be happening. The sense of speed eliminating distance stimulated pre-existing fears that local cultures were being lost in a long process of Tokyo-centric homogenization that had been under way since the
early twentieth century, spurred by continuous advances in technologies
of transportation and communication, especially the recent advent of television. For example, in the 1959 *Yomiuri*-sponsored discussion of the future of Tokyo over the coming decade, former Waseda University president
Shimada Kōichi, an economist whose research focused on transportation,

noted the rising centralization of Japanese culture "on all fronts," pointing out that a large majority of the content of radio and television broadcasts was made in Tokyo.[92] New infrastructures for high-speed travel on road and rail threatened to exacerbate this trend.

The fear of homogenization is one example of the ways in which the spatial shrinking and linking effects of the bullet train, though similar to those of the nineteenth and early twentieth century, were particular to their times. The unmatched speed, the dramatic increase in transportation capacity (especially in combination with new expressways), the possibility of a day trip between Japan's two largest cities, and excitement about the technology and design of the train itself all contributed to a sense of wonder and heightened connection. The revision of schedules on connecting lines and emerging plans to add high-speed rail alongside other trunk lines gave this single new train an immediate national impact, as opposed to the gradual expansion of rail over several decades following the opening of the first line in 1872 and later across the empire. And while the railroad's dampening of regional difference was celebrated in the earlier era, in part because it contributed to the government's nation-building (and to some extent imperialist) goals, in the context of the homogenization of culture by mass media in the mid-twentieth century, this facet of high-speed rail felt like a threat to local differences.[93]

[margin handwriting: Common theme of dream becoming reality and turning to a nightmare]

INFRASTRUCTURE AND (OUTER) SPACE

In 1965, a giant turtle-like monster, having slept beneath the Arctic ice for untold ages, was accidentally awakened by a nuclear explosion resulting from a minor Cold War skirmish on the Japanese silver screen. This ancient creature, Gamera, makes its way to Japan and begins to demolish the symbols of the modern Japanese city. In its search for sustenance, Gamera destroys a bullet train, a geothermal power plant, and an unfortunate swath of Tokyo. In synecdochic symbolism, it spends a fair portion of its rampage of the capital demolishing a building identified in prominent English-language signage by the name "New Tokyo." The movie poster (fig. 2) highlights the bullet train as one of the three symbols of contemporary Tokyo under attack: with a crumpled Tokyo Tower and a cracked Diet building in the background, Gamera grasps the front cars in one hand while the rest of the train tumbles to the ground from its elevated track. Tokyo, and the

FIGURE 2. Movie poster for *Gamera, the Giant Monster* (*Dai kaijū Gamera*, 1965).

modern world as a whole, is saved from this menacing metaphor only by international cooperation among the world's major technological powers: the United States and the Soviet Union, with Japan playing the role of intermediary. In the end, the destruction of Japanese space by the reptilian symbol of the nuclear threat is countered by deft social construction of a new space: a rocket, the transportation technology of the future, sends Gamera into outer space.[94]

The film *Gamera, the Giant Monster* actively reimagines the impact of transportation technologies on the spaces of the Cold War era: the Arctic becomes a site of battle between nuclear-armed jet fighters, and rockets make outer space a venue for superpower cooperation. But governmental and scientific institutions do not have a lock on the social construction of those spaces. A young boy who had insisted throughout the film on the monster's inherent good nature vows to go someday to Mars to visit the gigantic fire-breathing turtle in its new home. The film is a critique of atomic weapons and Cold War rivalry, not transportation technology, but the function of the fictional rocket echoes what the real bullet train reveals: the wielders of institutional power can construct space through the creation of transportation technology, but others are free to reinvent that space according to their own ideas and perceptions.

Planners and scholars conceived the space of the bullet train line in terms of their particular goals. In 1960s Japan, this meant promoting economic growth at all costs. Officials at the national and local levels focused on bolstering economic development and redistributing population by accelerating flows of goods and people between and beyond the cities of the Tōkaidō. But passengers, observers, and producers of popular culture made their own contributions to the social construction of the space, bending it to their particular uses and perceptions. These had to do not primarily with economic transactions, but rather with culture and community. And the new line exerted its own agency in the social construction of space; its speed and aesthetics inspired both actual and imagined excursions, expanding the range of the familiar by engendering new patterns of mobility and new uses of space, which reoriented social relations and people's relationships to their cities and infrastructures. Producers of popular culture pushed against rosy views by creating alternative images and understandings, using the bullet train as a symbol to highlight the problems of the high-speed growth economy: the underlying corruption, the people left behind, the damage done to lives and landscapes. In that sense, the new line embodied and amplified one of the primary tensions in postwar Japanese society. Images of the bullet train in popular culture highlighted the contest between unfettered development and the preservation of tradition and nature, the challenges of maintaining cultural diversity in an era of mass media and high-speed transportation, and the struggles of people swept aside or crushed by the juggernaut of economic growth driven by industrial expansion.

The question of the bullet train's economic role itself became a point of debate in competing views of Japan's future in light of national socioeconomic changes of the 1960s and 1970s. The next chapter will broaden the view from the Tōkaidō region to the nation as a whole to consider these issues in terms of the idea that Japan was becoming an "information society." The struggles of planners to use infrastructure as a tool of urban policy in the Tōkaidō region was echoed on the national level, as debates over the economic, social, and political impacts of the new line reverberated through the years of its construction, early operation, and subsequent expansion in the context of the social and economic changes that were dubbed the "informationization" of Japan. The positions staked out in the discourse on the Tōkaidō region—whether the new line would be a positive force bringing people together or a nefarious agent of homogenization and centralization—were echoed in discussions about expansion of the system and its reinterpretation as an infrastructure of information society. The inception of the bullet train was figured as a turning point for the railway industry as a whole and a marker of possible new directions for the national economy and society. Development plans, scholarly critiques of policy, and popular-culture representations of the process presented conflicting views of the meanings of high-speed rail for Japan's future.

Who gets access to information?
Is it fair?
Quickly moving into isn't always good

3 Railroad for the Information Society

THOUGH THE DEVELOPMENT of both automobile and air transportation in the early postwar decades led many to see railways as a sunset industry, the bullet train seemed to transform it from a relic of the past into a harbinger of the future. This change was propelled by the new train's sleek design, computerized controls, and smooth high-speed operation. But it also stemmed, in part, from its connection to developments in the basic patterns of the Japanese economy and society that scholars and policy makers were just beginning to notice at the time. Over the course of the 1960s, the idea took hold that the nation was in the midst of fundamental changes equivalent to a second industrial revolution. The emerging socioeconomic structure, variously referred to as "information industry society" (*jōhō sangyō shakai*), "information society" (*jōhō shakai*) or "informationized society" (*jōhōka shakai*), centered primarily on the growing prominence of information industries and communication technologies within the Japanese economy and society. The bullet train both marked and embodied this turning point, participating in the conceptualization of Japan as an information society. Examining the ways that two very different groups—urban planners and fiction writers—imagined the bullet train within the context of such socioeconomic transformations shows how information came to be a lens through which infrastructure was read and how the bullet train became a key for understanding social change.

As a source of global attention and national pride, the new line was drawn into evolving visions of Japan's future.[1] Though it had been conceived as the solution to problems of the industrial economy, it was soon reinterpreted as an essential element of post-industrial Japan. As a means of high-speed transportation, the train would move information (the knowledge held in travelers' minds and the materials in their briefcases) faster than the vehicles of the industrial era. And as a system developed in the computer age, the bullet train also relied on electronic communication and control, information moving almost at the speed of light. For these reasons, the new line became part of the "informationization" of Japan and has been called a "vanguard of the information society."[2] In some ways, this was an even more heroic narrative than JNR's own representation of the bullet train, placing it at the center of a reimagination of Japan that put the nation at the forefront of a new stage of economic and technological development. But that hero could appear as a villain when the informationization of society was defined as detrimental to the nation.

It may seem strange to consider a train as a means of communication at a time when electronic telecommunications technologies had already revolutionized long-distance communication. Beginning with the telegraph in the nineteenth century, and then intensifying with the growing availability of radios, telephones, and televisions in Japan by the 1960s, the capacity to transmit information in real time over long distances did not seem to leave much of a role in communications for a train, even one that could go from one end of the Tōkaidō to the other in just a few hours. And certainly information was not an explicit focus for the decision makers and planners at JNR. They worried about transport capacity and the bottlenecks that made for industrial inefficiencies, so their attention was on goods and people, not the information they carried. The role of the train in information flows, however, was important in helping people plan for and make sense of the changes taking place around them.

Within the early discourse of the information society, the bullet train—as a new, faster connection between distant places—was drawn into efforts to shape Japan's future by restructuring and expanding existing systems for the creation and dissemination of information. Examining the historical meanings attached to the new line in that context shows how urban planners and other intellectuals grappled with momentous changes taking place around them and articulated their visions for Japan's future. The intellectual discourse that placed the bullet train at the center of Japan's information society reveals a fundamental conflict between competing visions of what that future should look like. The debate over how to use the communications potential of high-speed rail as a tool of social and economic policy reflects a tension between planners' goals of maintaining social control and promoting individual freedom and equality. Langdon Winner highlights contrasting perspectives about the relationship between infrastructures and political forms. On the one hand, he explains, in the same year as the bullet train began operating, Lewis Mumford defined system-centered technologies (such as railroads) as essentially authoritarian, in contrast to more democratic human-centered ones. On the other hand, Winner notes, "Scarcely a new invention comes along that someone does not proclaim it the salvation of a free society."[3] Regarding the bullet train, pundits came down on both sides of this debate.

In an information society, control of information confers both power and wealth. The informationization of the bullet train by planners and fiction writers raised fundamental questions about building a rationalized society through centralized control versus fostering democracy by expanding freedom and choice. Their disparate views of the bullet train reflect different perspectives on information. Urban planners emphasized the new line's impact on information flows, considering how it could thus be used as a form of social control. They conceptualized the bullet train (along with other new technologies and infrastructures) as the central nervous system for the new information society, in which the central government in Tokyo acted as the brain, controlling the national body of Japan through expanding lines of communication. For example, soon after the bullet train opened, architect Tange Kenzō announced that the new line strengthened the information connections between Tokyo and Osaka such that "it can be called Japan's central nervous system, its spinal cord."[4]

While planners disagreed among themselves about how control of information could reshape society, challenges to official visions also came from materials of popular culture that mobilized the symbol of the bullet train as a mechanism for protest against the social ills they saw accompanying the informationization of Japan. Films and novels featuring the bullet train—such as Kajiyama Toshiyuki's best-selling detective novel about corruption in land acquisition for the "dream super-express"—used

uneven access to information as key plot points in fictional (and semifictional) stories of criminal activities and unfair dealings surrounding the
planning, construction, and early operation of the line. These stories presented a vision in sharp contrast to planners' image of fast-flowing information. Instead, they placed the bullet train at the crux of a society and
economy shaped by information flows that trickled or gushed or got diverted in unfair or illicit ways. Where planners saw expanded communication networks as a democratizing force, the fictional works examined here,
by showing how unequal access to information channels and sources translated into political, economic, and social inequalities, depict the growth of
an information society as reifying and even exacerbating inequality in both
economic opportunity and political power. This chapter examines the writings of three major figures in urban planning and three cultural representations of the bullet train to see how it was used in arguments for political,
social, and economic change.

Conceptualization of the bullet train in terms of information within urban planning was part of the public grappling with the challenges of the
1960s. The economic dislocation accompanying high-speed growth and the
transition from an industrial to an information society was only one part
of it. Japan was also in the midst of political and demographic crises. The
massive demonstrations against the 1960 revision of the U.S.-Japan Security Treaty (Nichi-Bei Anzen Hoshō Jōyaku, or Anpo) both reflected and
intensified a national struggle over the nature of Japan's postwar democracy. And the nation was beset by an intensification of the long-standing
twin problems of urban overpopulation and rural depopulation. In this
context, high-speed rail became a tool that might be used to solve these
problems. Popular fiction about the train provided a counternarrative to
planners' rosy views of free-flowing information undergirding a democratic and prosperous society. By highlighting blockages and inequalities in
access to information, stories of crime, corruption, and exploitation underscored the idealism of planners' depictions. Stories about the bullet train
thus contributed to national debates about democracy and economic development, casting doubt on the claims made by conservative leaders, but
in an alternative voice, perhaps less threatening to a mainstream audience
than leftist protesters, and therefore potentially achieving an impact that
other forms of protest could not.

THE IDEA OF AN INFORMATION SOCIETY

Anthropologist Umesao Tadao, one of the developers of the concept of information society, foresaw a fundamental transformation of Japanese society and economy and identified the operation of the bullet train as one of the many factors driving these changes. Umesao coined the phrase "information industry" (*jōhō sangyō*) in 1961, then elaborated on the concept in a 1963 article predicting the rise of the information industry, defined to include any enterprise dealing with information as a commodity, including its creation, storage, sale, and dissemination.[5] His definition of this new industrial category encompassed the mass media and publishing industries, as well as the construction and operation of communications infrastructures, and things like detective agencies and tourism, which he interpreted as the commodification of "experiential information." As a means of speeding up the transmission of information carried by people, whether physically recorded or stored only in their minds, the bullet train helped promote information industries and could itself be situated within that category. Originally published in a broadcasting industry journal, Umesao's ideas reached a broader audience when this article was reprinted just a few months later in the popular journal of opinion *Central Review* (*Chūō kōron*).[6] Although the specifics of his argument, which mapped the development of information society along the path of evolution in the animal kingdom, have been criticized as an explanation of social change, nevertheless, the article is important in the history of information society because it helped start the information society boom in Japan, a flood of books and articles that, by bringing the idea to the forefront of many disciplines, promoted the actual "informationization" of Japanese society.[7]

An extension of the idea of information industry, "informationized society" is a concept that gained traction in intellectual and policy planning circles in Japan by the late 1960s. Hayashi Yūjirō, a professor at the Tokyo Institute of Technology and advisor to the governmental Economic Planning Agency, began to use the concept in 1967 and published a book on the topic in 1969. In the same year, study groups attached to the Economic Planning Agency and the Industrial Structure Council (a think tank operating under the aegis of the Ministry of International Trade and Industry) produced official reports and plans grappling with this phenomenon.[8] Hayashi explained informationization as a process of social and economic change, analogous to nineteenth-century industrialization, by which value comes to be produced increasingly by intangible information, more than tangible material goods.[9] Part of that process was a trend toward viewing Japan's economy, industrial structure, and society as a whole in terms of information.

The bullet train provides an example of this change in the ways that scholars and policy makers viewed the economy. Though the new line was built to solve a problem impeding industrial growth—a transportation bottleneck along the country's most dense industrial corridor—Umesao placed it firmly within the category of information industry. The bullet train differed from earlier railroad systems, he explained, in that it carried no freight, but only passengers. In Umesao's view, this meant that "it carries information, or it carries people who are stuffed full of information. . . . Those who have meetings in Tokyo or Osaka or the like, it carries that kind of people."[10] Though he refers specifically here to business travelers using the bullet train to get to meetings, his earlier explanation that anyone bringing their travel experiences home with them should be seen as an information carrier meant that both business and leisure passengers were traveling for the purpose of transmitting information. And the significance of this particular characteristic of the new high-speed train extended beyond just one line to the entire technology of rail transportation. Umesao predicted that the bullet train was a harbinger of broader change: "The Shinkansen, essentially a system for handling information, is clearly beginning to show the transformation of the railway itself from a freight-transport system to an information-transport system."[11] In fact, planners at JNR had initially intended to run super-express freight trains on the new line at night, when passenger trains were not running, but that plan was scrapped because considerations of cost and track maintenance made it impracticable.[12] Though these practical reasons were behind the decision not to use the new line for freight, from Umesao's point of view, limiting the line to passenger traffic had a profound impact on its essential character. A transportation system conceived as the solution to an industrial problem became an infrastructure of the information society.

By the time the train actually opened, when the decision against freight had already been made, the idea of the bullet train as part of the information society had even seeped into the thinking of the JNR leadership.

In May 1964, soon after resigning as JNR president (to take responsibility for tremendous cost overruns on bullet train construction), Sogō Shinji gave a speech before the Japan Transportation Association (Nihon Kōtsū Kyōkai). He began by highlighting the birth and dominance of the railway in early industrial societies but then noted that, in response to challenges from other forms of ground transportation, it had become a "sunset industry." However, he argued, the ongoing second industrial revolution, in which machines are taking over brain functions, would give the railway an advantage over other forms of transportation, as rails were conducive to the technological revolution of automation. Sogō foresaw the information age as a new railway age. JNR's bullet train, he said, was the product of this technological revolution, citing the electronic communications that made high-speed operations possible and safe, including centralized traffic control of the entire line, on-train signaling, and automatic speed control and braking. "With all this," he concluded, "it is basically the world's first new railroad that widely adopts the recent technology revolution."[13] According to Sogō, this not only made a profound contribution to the world's railroad technologies but also created a central place for JNR in the new information-based Japanese economy as the nucleus of technological development.

In a retrospective view of the bullet train's impact, Umesao used the popular neural metaphor to emphasize its profound historical significance. As "a kind of information transmission system," he argued, the bullet train could be compared "not to the vascular system carrying nutrition through the entire body, but rather to the central nervous system that controls and coordinates all of its functions."[14] In the late nineteenth century, the nation-building potential of railways was often represented through images of "iron rails as the bones and sinews of the nation."[15] In addition, the sense that railroads carried the nourishment of agricultural products, industrial raw materials, and urban production throughout the "body" of the nation produced a vascular imagery and, eventually, the labeling of trunk lines as "arteries."[16] The shift from arterial to neural metaphors that marked discussions of the bullet train reflects a new conception of rail transportation, from the classic industrial model to an information society understanding, in which trains exercised a "neural" control function.[17]

Looking back on the era as he organized his writings for publication two decades later, Umesao considered the reason for this shift and argued that it reflected a fundamental change made by the bullet train to the nature of the railway. He suggested that the arterial metaphor had never completely captured the reality of railways, which had the dual function of carrying both goods and passengers. The train's involvement in human mobility made it a medium of communication, he wrote, because people move primarily for the purpose of acquiring or exchanging information, whether that is in the form of a business transaction, a religious pilgrimage (an information exchange between gods and humans), or tourism (experiential information). From that perspective, because the railroad had increasingly lost its freight transport function to trucks and other transport facilities and thus increased its communicative function, it should be considered "a major information facility of modern society."[18]

This was becoming even more true as modes of transportation were changing. In 1950, the railway carried 52 percent of all freight and 92 percent of passenger transport; a decade later, those percentages had fallen to 39 and 76 percent, respectively; by 1970, the numbers were down to 18 and 49 percent, while motor vehicles carried 39 percent of freight and just under half of all passengers.[19] Both passengers and goods were rapidly moving off the rails to other paths of transportation (primarily roads), but the railway's share of freight transportation was extremely low, even as trains maintained a sizable hold on passenger traffic. Looking to the future, Umesao held that, with the increase of shipping by truck and other methods, the railroad must emphasize tourism and other information activities. "The Shinkansen is steadily expanding in place of existing lines, but it does not carry freight. Passengers move carrying information or move in search of information. For the railway, the only path forward is through specialization as an information system."[20] Umesao essentially called for a transformation of the railroad industry to adapt to the overwhelming trends of the economy as a whole. And the bullet train was leading the way.

COMMUNICATION AND CONTROL ON THE TŌKAIDŌ

Umesao's neural metaphor became ubiquitous with the putative dawn of the information age, as pundits and planners assigned the functions of brain, spinal cord, and nerves to various aspects of the nation's infrastructure, reflecting the role of transportation and communication technologies in social control. The relationship between communication and control was a crucial factor in urban planners' treatment of infrastructures like the bullet train in terms of neural analogies. In higher-order organisms, the brain controls the body by means of messages sent via the spinal cord and nerves, making adjustments on the basis of information delivered back to the brain along the same path; in the same way, the bullet train was imagined as facilitating central control by delivering both command and feedback between the "brain" in Tokyo and the rest of the national body of Japan. This was related, in part, to specific aspects of the functioning of the bullet train system, especially the electronic centralized traffic control system, which integrated the monitoring, management, and guidance of bullet trains on the line through operation display panels at the headquarters in Tokyo Station and communication with drivers and crews on the trains. With increasing traffic, this system was upgraded to the computer-assisted COMTRAC (Computer-Aided Traffic Control) system in 1972.[21]

Imagination of the bullet train as part of a neural system was also influenced by the growing sway of cybernetics, a concept developed by American information studies pioneer Norbert Wiener during the 1940s and 1950s and soon popularized in Japan. Cybernetics, the study of the relationship between communication and control, proposed a basic analogy between machines and animals in their use of information-feedback paths for self-control: in both cases, the command center (brain) sends a command via the nervous system, information returns by the same path, and the brain responds by adjusting and sending a new command for action, in a constantly repeating pattern. Wiener argued that the feedback mechanisms that control the functions of machines were analogous not only to living organisms but also to society as a whole.[22] This certainly implied centralized control. In *The Human Use of Human Beings: Cybernetics and Society*, he wrote, "where a man's word goes, and where his power of perception goes, to that point his control and in a sense his physical existence is extended. To see and to give commands to the whole world is almost the same as being everywhere."[23] But for Wiener, this did not imply despotism. Rather, he viewed control of the means of communication as the glue that holds together any organism—including human society.[24] In the United States, cybernetics and information theory gained broad use in explanations of human behavior across many fields—including management, psychology, sociology, political science, and anthropology—and cybernetics systems modeling was prevalent in urban planning by the early 1960s.[25]

Several of Wiener's writings on cybernetics were translated into Jaɪ
the 1950s, and the major daily *Yomiuri shinbun* deemed the earlie
a "must-read."[26] Japanese scholars in many disciplines joined the
tional cybernetics boom. Urban planning, which was a relatively
cipline in Japan, was a fertile field for the burgeoning fascination with the
function and implications of information flows and feedback loops, especially in the area of communications and transportation infrastructure.

The intimate connection between communication and control made the
bullet train particularly interesting to planners looking for ways to solve Japan's problems in the context of informationization. They foresaw faster,
more efficient flows of information ameliorating two of the major challenges facing Japanese society in the 1960s: the crisis of democracy surrounding revision of the U.S.-Japan Security Treaty and the increasingly
uneven geographic distribution of population and industry. Optimistic
views held that greater information flows would mean expanded access to
information, a key element of a democratic society, and that increased mobility (of both people and information) would potentially slow the tide of
urbanization. However, critics pointed out that the informationization of
Japanese society seemed, in fact, to reinforce inequalities by simply transforming the elite of industrial society into an information elite with greater
access to and control of information, which they could use to defend their
status.

Japanese planners were not the first or last to think of transportation
and communications infrastructures in terms of control. From the roads
of the Roman Empire to internet protocols in our contemporary world, institutions and individuals have mobilized new information technologies to
expand or bolster their control over land and populations.[27] The general
path of Japan's first bullet train had been used in such ways since at least
the seventeenth century, when horsemen barreled down the Tōkaidō road,
conveying the intelligence and commands that facilitated shogunal governance. Connecting the shogun's capital in Edo (now Tokyo) with the imperial palace in Kyoto, this road was a central path for the flow of information
to and from the central government. Well before the introduction of railroad technology to Japan, the Tōkaidō was the most important of the five
main roads developed and maintained, but also closely monitored and regulated, by the Tokugawa shogunate as a means to establish and solidify its

control.[28] This channel was thus initially developed as part of a state effort to facilitate and control mobility and communication in order to maintain its power and extend its range of governance.

That method of solidifying and exerting power continued under the new regime that took over in 1868 with the Meiji emperor as its titular head. As part of a wide-ranging program of modernization along Western lines, the Meiji government built the infrastructure for an industrialized capitalist economy in Japan, including railroads. They prioritized the Tōkaidō Line, which was completed in 1889 and would continue to be Japan's busiest trunk railway line throughout the decades of industrialization, war, and reconstruction. That same process of industrialization also encouraged a population drain from rural to urban areas, which became a perennial matter of concern for urban planners.[29] In the emerging information society of 1960s Japan, the patterns of mobility created by the bullet train inspired new ideas about the potential for flows of information to reorganize the national space of Japan, in terms of population, industry, and economic activity. Urban planners linked the communication functions of the bullet train to central issues of regional and national development.

Planners are exemplified here by an architect, a politician, and a sociologist. Tange Kenzō was one of the most prominent Japanese architects of the time. After working for a Japanese architect in Manchuria in the 1930s, he studied at the University of Tokyo, where he became a professor in 1946 and led the so-called Tange Lab. In that position, he cultivated and maintained close personal and institutional ties to upper-level bureaucrats and government agencies. As is clear in his 1960 plan for Tokyo's development examined in Chapter 2, Tange extended his attention beyond discrete buildings themselves to consider how buildings connected to infrastructures and fit into larger urban and regional spaces. Tanaka Kakuei, a leader of the conservative ruling party, had served as minister of posts and telecommunications, then as minister of finance during the final years of bullet train construction, and as minister of international trade and industry before becoming prime minister in 1972. Isomura Eiichi was a pioneer in the field of urban sociology, having helped establish the discipline in Japan in the early postwar period. He spent twenty-five years working for the municipal government of Tokyo before embarking on a second career in academia in 1953. At the time of the bullet train's construction, he was a professor of urban sociology at Tokyo Metropolitan University, and in 1966 he moved to Tōyō University, where he later served as university president. As evidenced by his long participation in and leadership of the Social Integration Policy Council (later renamed the Regional Improvement Policy Council) and by his ideas about a Tōkaidō megalopolis discussed in the previous chapter, he was particularly concerned with improving human welfare through regional planning.

Seeing the new bullet train as a communications infrastructure, these men relied on the relationship between communication and control—control of population flows, of consumption and leisure activities, of economic transactions—to predict and analyze the social and economic impact of high-speed rail and make recommendations for urban and regional development. Though they all premised their work on the understanding that the redirection of information flows through the development of infrastructure was a means of social control, their disparate approaches resulted in overlapping but contrasting visions. Tange, writing in the early 1960s and inspired by Wiener's popular cybernetics model, promoted an efficient, rational social framework through a metaphor of society as an organic structure that relied on bullet trains and other information connections as feedback mechanisms that would allow the entire Japanese archipelago to operate as a single self-regulating entity. A decade later, as the concept of information society was becoming widely accepted—in connection not only with the growing prominence of information industries in the national and global economies, but also with the related development of the technologies of instant telecommunication—Tanaka (or those who wrote plans in his name) sought to control space by channeling information. His plan involved using the power of the national government to direct the flow of information (and hence, he argued, people and industry) to points decided in Tokyo by expanding the bullet train system and other communications infrastructures. In a critique of that plan, Isomura presented an opposing view of the effect of creating information connections with new high-speed rail lines. His findings challenged the utopian, teleological narrative of information society, which presented increasing flows of information as inherently and universally beneficial, or as progress toward a higher plane of democracy, rationality, and efficiency. He argued to the contrary that strong information connections such as bullet trains

ι to increase central power (i.e., an ability to exert control over sur-
ιing spaces) in ways that were not always desired by people outside the
r.

Ultimately, the ambitious plans proposed by Tange and Tanaka and critiqued by Isomura were not fully realized, but attention to their arguments about the future of the Japanese archipelago reveal the efforts of urban planners to grapple with changing views of the role of communication infrastructure in social control developed in the field of information studies. None of these three individuals was directly involved in the planning of the New Tōkaidō Line, but they represent one important set of perspectives on transportation infrastructures. All three tied dominant concepts of information and society to popular excitement about the bullet train and used these informationized interpretations of high-speed rail to shape new visions of Japan's cities, the Tōkaidō region, and the nation as a whole.

The bullet train ran straight down the center of Tange's vision of Japan's future. Because the new line would significantly expand the region's communications infrastructure, he saw it as an important factor helping to make the Tōkaidō belt into the "spinal cord" of the emerging information society.[30] Tange was among the first to interpret the bullet train in terms of cybernetics. That concept was fundamental to his early 1960s urban planning for Tokyo and the Tōkaidō area and more broadly to the architectural movement known as Metabolism, of which he was a central figure as the mentor of its founding members. The foundational principle of this movement, which was prominent in 1960s Japan, was to design buildings and cities with an organic adaptability to change; its proponents insisted that cities must be seen not as a set of fixed buildings and infrastructures but rather in terms of a constant process of renewal. The idea of urban space as living and growing was central for Tange, who held that effective spatial planning required consideration of Tokyo, the Tōkaidō region, and, indeed, the entire Japanese archipelago as a single organic structure.

Even before the label "information society" was popularized in Japan, Tange had begun to rethink urban planning in terms of information flows. Cybernetics had captured his imagination in the late 1950s and soon became the basis of his radical reconceptualization of nationwide urban development, based on the argument that, as an organic structure, society requires strong and capacious information connections to facilitate its self-control. It was during his 1959 stint as a visiting professor at the Massachusetts Institute of Technology (MIT) that he began to think about space in terms of information, communication, and cybernetics. MIT was a dynamic center of information studies at the time, boasting among its faculty several prominent figures in the development of information theory, including Wiener, the originator of cybernetics; Claude Shannon, often called the "father of the information age"; cybernetician Warren McCulloch; and logician Walter Pitts. Though it is not clear whether Tange had any direct involvement with this group of scholars, echoes of their ideas began to appear in his work as soon as he returned to Japan the following year. He later pinpointed his stay at MIT as the beginning of his interest in information. In a 1985 recollection, he stated:

> At the time, the term *informational society* had not come into use; but I was aware of the importance of communication and information. . . . I predicted that, in the coming years, space would have to be a realm for the communication of information and that communications space networks would create both architecture and urban spatial structures. I felt that a new movement from functionalism [the architectural principle of industrial society] to structuralism was necessary. While at MIT, I pondered this problem while walking about and looking at aerial photograph[s] of Tokyo.[31]

Tange began applying the concepts of cybernetics to urban and regional planning with the ambitious "1960 Plan for Tokyo," prepared in cooperation with his research group and drawing on work he did at MIT. In this design, introduced to the Japanese public through a televised New Year's Day presentation, Tange used Tokyo's population milestone of ten million people as a prompt to tackle the specific problems created by urban overpopulation. Concerned primarily with facilitating mobility between and through cities that were becoming more crowded and sprawling every year, he focused on designing efficient communications and transportation infrastructures. Though the plan did not use the word "information," it emphasized the central role of communication in the modern megacity and contemporary society as a whole. As Rem Koolhaas and his co-authors explain in their extensive exploration of Tange and the Metabolists: "What Tange's team does not mention is industry: their city is postindustrial,

ned for the tertiary sector, a city composed by its flows of communi-
n, information, and road traffic."³² Though no one had yet called Ja-
_ an information society, Tange's plans were based on the developments
implied by that label. The plan stated, "The process of economic circulation within a given country is determined by a complicated system of relationships, in which government, politics, finance, control of production
and consumption, technology, and communications are all intimately and
mutually linked." And while he mentioned "production and consumption,"
this does not imply a focus on industry. Rather, he noted the process by
which industry in Japan was being replaced by what he called "organiza-
tional activity"—but could equally well be labeled "information"—which
"decides everything, creating wisdom, producing values, and connecting
them with the world."³³ Tange had already proposed methods to ease population pressure on Tokyo through communications infrastructure during
the Asia-Pacific War, with his prize-winning but never realized proposal
for a Greater East Asia memorial, which envisioned an urbanized axis linking Tokyo to the foot of Mount Fuji.³⁴ Perhaps a glimmer of the future information society was already visible two decades earlier.

Not only Tokyo but the entire country was struggling with demographic change as populations shifted from rural to urban areas, especially
the cities of the Tōkaidō region. Between 1950 and 1960, the national population increased by about ten million people, but urban areas increased by
roughly twenty-eight million, while rural areas fell by eighteen million.³⁵
Responding to this broad demographic challenge, Tange further developed
the concepts laid out in the 1960 Plan, expanding the scope from metrop-
olis to archipelago with a proposal for a "national axis" made up of communications and transportation infrastructures, a new spinal cord for a
restructured national body. In this conceptualization of a "Tōkaidō Megalopolis," Tange for the first time made his focus on information explicit. He
saw the ongoing "information revolution" identified by Umesao as the fundamental change shaping Japan's future. "If the first industrial revolution
was a revolution connected to the production of goods, one can say that
the second industrial revolution is a revolution in handling information.
As opposed to a revolution of the body, one can say it is a revolution of the
nerves and brain. By reinforcing the nervous system in this way, contemporary society is developing its organic structure."³⁶ Just weeks before the

opening of the New Tōkaidō Line, in August 1964, Tange incorporated this new infrastructure into a plan for the future organization of the Japanese archipelago, centered on development of a Tōkaidō megalopolis, a continuous urban space connecting Tokyo and Osaka and eventually extending from one end of Japan to the other.[37]

In this plan, Tange applied Wiener's cybernetic theory to emerging concepts of Japan as an information society by pairing dynamic economic and demographic changes with what he saw as a deepening of Japanese society's organic structure and linking both to information.[38] His premise was that spatial structures must be able to grow and change, that they must be divided by function, and that these functions must be coordinated and aggregated to support the life of the organic structure (i.e., society). Based on the cybernetic analogy that machines, living beings, and society as a whole all relied on feedback mechanisms for self-control, he emphasized the importance of "informational couplings" that allowed for two-way communication: command and feedback. In organisms, the nervous system provides such a feedback mechanism between separate functions, and in social organizations, that work is done by communication infrastructures. Tange concluded, therefore, that the development of communication technology was effectively deepening the organic structure of society, such that it could automatically control its actions and adapt to changing circumstances better than in the older industrial society.[39]

In spite of the development of telecommunications technologies, Tange emphasized high-speed rail and roads. The instruments of electronic mass communication, such as radio and television, contributed to the flow of information, but they lacked the essential element of feedback. Although the telephone allowed two-way communication over long distances, Tange dismissed such "indirect communication" as insufficiently developed to replace face-to-face exchanges among many people. He argued, therefore, that transportation was becoming increasingly important as a means of information connection. Thus, for Tange, the two mutually reinforcing processes characterizing contemporary Japan (and the capitalist world as a whole)— the intensification of economic and population growth and the deepening of the organic structure of society—both relied on increasingly fast and efficient flows of information embodied primarily in human transportation. Comparing urban areas to artificial intelligence devices, he stated, "Just as every electron in an electronic brain takes up and moves information, so information processing takes place through human transportation, as every individual person moves around like an electron."[40] In the anthropomorphized metabolic structure of the Japanese national body, Tange positioned people as electrons, whirling around in the brain (Tokyo) and moving along lines of the network to other parts of the organism with messages from the brain.[41] This analogy led him to conclude that the future organic society needed a new physical structure: a single megalopolis that would ultimately extend over the entire archipelago, with peripheral cities connected all along the length of the central "spinal column." Tange's was fundamentally a centralizing plan. By positing one city as the brain of an organically structured society, he envisioned information flows as enhancing central control of the nation as a whole.

The megalopolis concept reveals a contradiction inherent in Tange's work, and in the Metabolist movement more generally, between the element of control implied in centrally planned designs and the aim of designing for greater freedom and equality in the aftermath of the political crisis surrounding the 1960 revision of Japan's security treaty with the United States. While the emphasis on creating more efficient flows of information accommodated democratizing principles, scholars have identified autocratic elements in Tange's work. For instance, architecture critic Hajime Yatsuka and media scholar Yuriko Furuhata both draw a connection between Tange's plans for regional development and the biopolitical spatial imagination behind colonial planning for Manchuria in the 1930s and early 1940s.[42] Architectural historian Zhongjie Lin notes a tension not only among the Metabolists but in modernity itself between the competing ideologies of rationalism and libertarianism, producing conflicting ideals of society as either highly regulated and centralized or decentralized and locally organized. He sees in Tange's approach to regional planning the architect's "technocratic ambition, that is, to control the development of the whole nation by means of modern technology, management, and planning."[43] But he argues that this approach—the Metabolists' desire for a master planner—contradicted their fundamental goal of designing urban spaces that would cultivate democratic societies. Furuhata shows that Japanese scholars of communication studies were very much aware of the tension between the "emancipatory potential of communication" and the possibility that it might become, as liberal intellectual Tsurumi Shunsuke put it, "a technique that benefits the oppressor."[44]

That contradiction is visible in Tange's plan, aimed at increasing the quantity, speed, and efficiency of the information transfer that is central to both a vibrant economy and a healthy democracy, but through highly centralized planning. Tange himself claimed to be designing for "an atmosphere of freedom and constant choice" through the development of information connections.[45] But, in accordance with the cybernetic analogy, he proposed a structure of completely centralized control by a brain center in Tokyo, which sent out commands and received feedback from all parts of the nation through a system of high-speed trains and expressways. The cybernetic approach to urban planning helped smooth this contradiction between control and design for greater freedom by placing autonomous individuals within a centrally planned information feedback loop. The biological analogy helped naturalize central control by likening it to the essential function of control of the body by the brain.[46]

The tension between freedom and control was central to the clash between two main views of the functions of the bullet train in an information society that emerged nearly a decade later. In 1972, the construction of a national network of high-speed rail and other infrastructures of transportation and communication was at the center of a plan for rebuilding the Japanese archipelago in the context of informationization. The much-discussed book *Nihon rettō kaizō ron* (translated into English as *Building a New Japan: A Plan for Remodeling the Japanese Archipelago*) was published under Tanaka Kakuei's name on the eve of his election as prime minister, eight years after the New Tōkaidō Line opened and Tange presented his megalopolitan vision.[47] Though the extent of Tanaka's actual involvement in the plan's creation or commitment to its ideas is in doubt, because it was attributed to and claimed by him, the plan was and remains connected to him in the public mind.[48] Urban sociologist Isomura Eiichi critiqued the Tanaka plan by attacking its basic assumptions about the impact of the bullet train as a carrier of information on population distribution, urban development, and the decentralization of power.

By this time, the idea of Japan as an information society had made its way into governing circles and attained the status of official policy, standing at the center of a number of government plans and reports.[49] For example, the Economic Planning Agency's 1969 New Comprehensive National Development Plan took as its basic premise that Japan was becoming an information society as the result of "rapid progress in the so-called 'Second Industrial Revolution' based on the information revolution, internationalization, and technological innovation."[50] Policies for land utilization, regional development, and infrastructural expansion, including construction of new high-speed railways and other transportation and communications networks, were all built around the interconnected goals of furthering and adapting to this new reality. Elements of this blueprint for development are reminiscent of Tange's proposals, especially its understanding of the integrating functions of these expanding networks, which it envisions as "the main axis of Japan," facilitating "the central management function of Tokyo."[51] Echoes of Tange's ideas are perhaps not surprising, given that two graduates of Tange Lab, Shimokōbe Atsushi and Obayashi Jun'ichirō, helped shape the government's development plans throughout the 1960s as officials in the Economic Planning Agency. Shimokōbe was an especially significant figure in the upper levels of the government bureaucracy and from that position played the role of "puppet master" for the Metabolist group, implementing its vision from behind the scenes.[52]

Though he later denied any direct involvement, Shimokōbe is said also to have exerted a strong influence on *Building a New Japan*.[53] The book was not revolutionary in its main ideas, bearing a heavy similarity to the Economic Planning Agency's 1969 development plan. The promotion of a nationwide system of high-speed rail, in particular, had already won official support in the 1970 Nationwide Shinkansen Railway Development Act (*Zenkoku Shinkansen tetsudō seibi hō*), and construction of a line extending west from Osaka to Hakata was under way. But the book's putative authorship by a powerful and charismatic politician stimulated popular interest, making it a best-seller.[54] In a 2005 interview, Shimokōbe cast doubt onto Tanaka's own commitment to the ideas it contained.[55] But the question of whether Tanaka truly supported the plan as a whole (or had any significant role in its creation) is beside the point. The energetic response to the book—both positive and negative—among scholars and in the media, along with its purported authorship by the nation's top political figure, show the importance of the ideas it contained.

Tanaka's plan shared the same goal as Tange's—ameliorating the congestion and other problems caused by excessive concentration of population

and industry in the Tokyo-Nagoya-Osaka belt—and started from the sa
premise that improving flows of information would contribute to pub
lic welfare. This grand proposal for massive infrastructural development
across the nation included the expansion of information connections as
part of central government efforts to shape space and mobility, arguing that
the construction of new bullet train lines and other information infrastructures could be used to redistribute population and industry more evenly
across the nation. Created at a moment when the communications potential
of computers was beginning to come into focus for policy makers, Tanaka's
book placed more emphasis than Tange had on telecommunications and
computing networks as the infrastructures of the information society, but
the nationwide expansion of the bullet train system was a major component
of this plan to develop information industries, redistribute population, and
decentralize power by making information more easily, quickly, and abun-
dantly available in peripheral cities. Like Tange, Tanaka argued that the
problems of uneven population distribution, as well as related regional economic disparities, could be eliminated through infrastructural planning.

In spite of the basic similarities between them, shaped by the specific
historical context of its creation and a different understanding of the impact of information flows on population mobility, the Tanaka plan's conception of the space to be created by newly built infrastructure differs dras-
tically from Tange's. Tange sought to speed up and expand connections
along the Tōkaidō; in contrast, Tanaka aimed to redirect those flows outward in order to redistribute both population and industry. Blaming "psy-
chological distance and the information gap" for preventing the regional
dispersion of industrial production and population, Tanaka concluded
that the construction of bullet train lines, expressways, and other networks
would solve the population problem, because giving people in regional cities equal access to information would weaken the incentives for them to
leave. Ultimately, the "focal cities" created by thousands of miles of super-
express railroads crisscrossing the country would be within a few hours of
each other, effectively unifying the urban centers and making them all as
accessible as suburbs of Tokyo.[56] While the bullet train alone would not effect these changes, it was one of the primary components of this plan.

While Tange envisioned the natural growth of a self-regulating organic
body, shaped by information feedback, Tanaka followed more of a Dr. Fran-
kenstein approach: information flowing along newly created paths would

act as an electric charge intended to reanimate declining areas on the periphery. This difference emerged from their particular attitudes and goals. Tange took as his starting point the understanding that people wanted to live in Tokyo (and other Tōkaidō cities) and, therefore, aimed to develop systems and structures to facilitate mobility and communication within, between, and beyond these heavily populated cities. Tanaka, on the other hand, as the political representative of one of the peripheral areas in question, had strong professional and personal incentives to promote a model that would pull people and industry out from the center.[57] It is perhaps no coincidence, then, that his plan aimed to direct information flows outward, so that new information industries would develop in rural areas and regional cities and people would, therefore, want to stay there. While Tange took collective personal preferences as a starting point, Tanaka hoped to use the power of information flows to change those preferences: a national network of super-express railways, along with improvement and expansion of roads, airports, ports, and computer networks, would be aimed at bringing all of the major areas of the country within one day's travel. This expansion of mobility, Tanaka held, would increase production and expand human activity, "including consumption, information transmittal, and recreation, enhancing all the functions of our society."[58] He envisioned new bullet train lines, in particular, as tools for regional development, "pump-priming investments" meant to promote sparsely populated areas and close the gap between those areas and the more industrialized, crowded, and prosperous Tōkaidō region.[59]

Though Tanaka did not share Tange's vision of an organic society, he used a similar neural metaphor, calling Tokyo and other major cities the "nerve centers" of political, economic, social, and cultural functions of Japanese life, a position that had been intensified in the information age. But, shifting to other anatomical metaphors to express breakdowns in the national body, he decried this structure as one that worked to subordinate and homogenize peripheral cities, while large cities suffered from "hypertrophy" and the "arteriosclerosis" of congested traffic, as well as "asthma" caused by the resulting pollution. Smaller regional cities, in contrast, relying on distant nerve centers with none of their own, are "like a human body without a brain."[60] His proposed remedy for these ills was to promote the development of knowledge-intensive industries in regional centers, so that

people would be lured away from Tokyo and other large cities, alo
tended national network of high-speed roads and rail.[61] In this re
plan relied on the assumption that speed and convenience of tra
encourage people to move out of Tokyo, rather than exacerbat
urbanization.

Tanaka also differed from Tange on the notion of centralized control,
arguing that "it is often erroneously asserted that the information society
is necessarily a centralized society. On the contrary, it is precisely because
of improved information systems that the regulatory powers of the national
government can and should be more decentralized and local governments
can and should be given an even freer hand in planning on the basis of
as much information as is available to the national government."[62] Tanaka
sought to control population mobility by channeling information flows, but
he claimed that this would ultimately decentralize control. These two disagreements between Tange and Tanaka highlight fundamental questions
for planners: can they control mobility by redirecting information flows,
and what impact would that have on centralized control over society?

Such questions were addressed by one prominent critic of the plan, Isomura Eiichi, then a professor of urban sociology at Tōyō University. Isomura responded to Tanaka's plan by challenging its underlying assumptions about the relationships among information flows, mobility, and power
and making a very different forecast of the impact that the expanding bullet
train network would have on the human decisions that cumulatively shaped
population distribution, economic activity, and local and national culture.
He predicted that, contrary to official expectations, expansion of the bullet
train system would not only worsen Japan's demographic problems but also
increase both the centralization of government administration and cultural
homogenization, a matter of concern since before the first line was built.[63]
Isomura was a visiting scholar at Harvard University in 1957, two years before Tange's stay at MIT, when these two universities together constituted a
hotbed of information studies, and he similarly brought ideas about information flows to the center of his analysis of the impact of high-speed rail on
Japanese society.[64] In November 1972, just months after the publication of
Tanaka's "remodeling" plan, he published an article identifying a different
kind of spatial reorganization resulting from the debut of high-speed rail.
Beginning with the fact that Tokyo, Osaka, and Nagoya had continued to

outpace outlying regions in population and economic growth, he explained that in the Tōkaidō region, where information channels were highly developed, the bullet train did indeed affect people and economic activity because, in general, information stimulates the receiver to action, and the bullet train gave people greater opportunity to act by increasing mobility. However, he wrote, rather than easing pressure on the capital, the new line "quickened the flow of brain power and the collection of information in Tokyo at the expense of all cities in the megalopolis along [its] route, including Osaka."[65] By highlighting actual demographic trends, Isomura disputed the notion that policy makers could control space and mobility as they expected through creation of more and faster information connections.

Isomura questioned Tanaka's reasoning that in an "informationized society" such as Japan's, "as long as the people have easy access to information, they will not feel the need to venture into the city and will be content to remain at home in the country," taking issue with its underlying assumptions about human nature and suggesting that people were more likely to have the opposite reaction. "The more developed a society's information system," he wrote, "the more people tend to gather at the source of information. In Japan, this source is composed of the Diet and the national government. Therefore, the question is not simply whether or not Tanaka's new cities can be built, but whether these cities can reduce the convergence of people on Tokyo as long as the government remains there."[66] Isomura concluded that it was not sufficient to expand and accelerate the conduits of information, but rather that the source of information itself must be moved. Therefore, he proposed dispersing the structure of government administration—a primary source of information—by building a new government center a short distance away at the foot of Mount Fuji.[67] His choice of location echoed Tange's wartime proposal for the Greater East Asia memorial but took it one step further by calling for not just a spiritual and cultural center but even the functions of government to be located at this historically and culturally meaningful space about 100 kilometers southwest of Tokyo.

Continuing his exploration of the bullet train's impact on space in a 1973 article on urban planning, Isomura acknowledged that although policy makers could not easily redistribute population, the weight of both bullet trains and expressways in regional development and urban planning gave the government a great deal of power to shape national, regional, and local space.[68] He also viewed the relationship between information and control differently than either Tange or Tanaka. Examining the impact of bullet train construction on urban and regional development, as well as on the perceptions, expectations, and fears of people in smaller regional cities slated for future bullet train stations, Isomura echoed Tange and Tanaka in placing the bullet train at the center of the emerging information society. But more than a plan for the future, his conclusions were a warning of things to come. He found that the consciousness of the Japanese people was becoming homogenized by the informationization of society and that the bullet train bore significant responsibility for cultural, economic, political, and administrative centralization. Directly contradicting Tanaka's claims that high-speed rail would contribute to the decentralization of power, Isomura argued that the structure of the emerging bullet train system made Tokyo like a main office of a geographically dispersed company, while regional cities were relegated to the position of branch offices, or even subsidiaries, communicating with the center via high-speed rail.[69] Already, in a 1969 study, economist Sanuki Toshio found that the bullet train had indeed contributed to an increase in the concentration of central management functions in Tokyo.[70]

The relationship between information and control was central to Isomura's understanding of the bullet train's impact on all aspects of social life. In his analysis, in an information society, the outward flow of information gave the providing city cultural, economic, and administrative influence over those receiving it, and high-speed rail was an important conduit for centrally produced and disseminated information. As the capital, Tokyo was the nucleus of Japan's information society, and Isomura noted that it therefore held the supreme position, one step higher than Osaka. By the same token, cities on the fastest Hikari super-express would place higher in the evolving urban hierarchy than those with stops on the slower Kodama bullet trains that stopped at each station. By the time he wrote this 1973 article, the second bullet train was partially open on the San'yō Line, extending the high-speed rail system from Tokyo west to Okayama. Isomura saw that Osaka's place at the junction of the Tōkaidō and San'yō bullet train lines boosted its regional clout, since "being within reach of Tokyo as a day trip further solidified its power base in relation to its periphery" in

1 Japan.[71] And to the extent that the Tōkaidō bullet train allowed To-
1 Osaka to operate as a single urban sphere, it increased control over
         ral areas by the Tōkaidō megalopolis as a whole. He even suggested
elsewhere that, as the bullet train increased communication between Tokyo and Osaka, intermediate cities would need a "defense system" against
terminal-city influence, in terms of both the potential obliteration of local
lifestyles and the erosion of local administrative autonomy due to greater
central control.[72] This work challenged optimistic predictions of the decentralizing effects of high-speed rail networks by considering how the flood
of incoming information threatened to overwhelm local cultures and local
governance. Isomura's work can be seen as an effort to salvage local autonomy and cultural diversity from the intensification of centralized control
accompanying the expansion of communications networks.

Isomura's argument highlights a characteristic that Tange's and Tanaka's
plans, as well as that of the Economic Planning Agency, have in common:
they are all examples of what historian Tessa Morris-Suzuki disparaged as
"a technocrats' Utopia, a public justification for the policies desired by economically powerful groups."[73] More recently, Shunya Yoshimi reinforced
this point, asserting that "information society theory fails to give due attention to the potentially conservative effects of quantitative changes, simply reinforcing the existing social order."[74] These plans assert that their goal
is to improve the welfare of society as a whole by restructuring information
flows, but they maintain the existing social hierarchies and structures.[75] As
is clear from the writings of advocates such as Tange and Tanaka, planners
had high hopes about the informationization of Japan, based on optimistic
assumptions that it would bring economic vitality, rational social organization, and other solutions to vital problems confronting Japanese society. But
those analyzing the social impact of the government's plan for information-
centric development warned that it might bring unintended negative effects,
such as increasing centralization of both administrative and cultural functions and perpetuation of economic and social inequalities.

CRITIQUES OF INFORMATION INEQUALITY

While Isomura presented reasoned critiques of the government's approach
to national development in academic venues, writers of fiction used representations of the bullet train to bring popular attention to sectors of the

population for whom the promise of the information society seemed to an empty one. Art and culture were active sites of political protest in years during which the bullet train was being planned and built. Historian William Marotti argues that in the late 1950s and early 1960s, cultural production took on new importance as a site of protest, as more mainstream structures of protest were co-opted or weakened by the conservative ruling party's success in connecting its governance with rapid economic growth, which undercut demands for social change, including greater democratic equality. Marotti examines the work of artists who resisted what they saw as "hidden forms of domination in the everyday world" through their art. In this way, cultural producers sought to disrupt the complacent acceptance of prosperity in place of democracy in "a demand for equality and real, instead of ersatz, popular sovereignty."[76]

Marotti highlights avant-garde artists, but mainstream cultural producers similarly used their work (and their privileged positions) to make political statements or press for change by highlighting social ills. For instance, two writers of fictional stories centering on the bullet train took up the effort of resisting the powerful and advocating real equality by representing back-room political and economic transactions surrounding the planning of the new line and by juxtaposing the advantages and imagined dangers inherent in its communications-dependent automated systems. Considering the treatment of information in a novel and two films— Kajiyama Toshiyuki's 1963 detective novel *Dream Super-Express* (*Yume no chōtokkyū*) and its film adaptation released the following year, titled *Black Super-Express* (*Kuro no chōtokkyū*), as well as the 1975 action film *Shinkansen Explosion* (*Shinkansen daibakuha*, released in the United States as *Bullet Train*)—demonstrates three different ways in which producers of popular culture took advantage of interest in the bullet train to convey a message.[77] First, they used the new line as an attention-grabbing symbol to highlight the potential problems created by Japan's transition to an information society, emphasizing the ways in which the new high-speed rail system both produced and conveyed information and the consequences of this function for individuals in various walks of life. Second, in contrast to the schemes of planners like Tange and Tanaka, in which efficient flows of information are the foundation of economic prosperity and public welfare, popular culture materials highlight the damage done by failures of communication and misinformation, such as the rumors and forecasts behind speculation, shady real estate dealings, and even violent crime. And third, while planners' use of high-speed rail to solve Japan's social problems was based on the promise of efficiency, productivity, freedom, and equality produced by increased flows of information, the darker view of the bullet train seen in these films and novels stems from the reality of uneven access to information.

In her 1988 study of the information society that was still taking shape in Japan at the time, Morris-Suzuki noted the potential problems that 1960s intellectuals saw accompanying the social and economic changes they described and predicted. The greater availability and more efficient flow of information contained the risk that information technologies might inadvertently work to erode the privacy of individuals and organizations. The information society might also be a severely stratified one, with a vast chasm between the "information elite" who create, control, and have access to a vast range of information, and the relatively powerless masses who consume only the information provided to them and are unable to obtain more exclusive or valuable information. And, as with any significant economic change, there was a danger of unemployment and other forms of economic dislocation resulting from shifts in the economy.[78]

The stories examined here address all of these problems, the issue of inequality most extensively. In this way, they presented a critique of the technocrats' utopia that the bullet train was helping to build and a contrasting perspective on the role of infrastructure in an information society. At the same time, novels and films delivered to a popular audience a sense of the growing importance of information in Japan's society and economy. The action of Kajiyama's novel takes place when the new line is still in the planning stages, not yet performing a communicative role. Therefore, it depicts the train not primarily as a carrier of information but rather as a catalyst for the creation, acquisition, and use of information with both economic and political value. This representation highlights a key aspect of information society: the fact that information itself produces value. Information studies scholar Hayashi Yūjirō explained this change in 1967. Defining an information society as one in which information produces value, he compares that to the previous era: "Regarding industrial society, one can say that it is an era when tangible material goods produce value; in contrast, in an information society, the information that produces value is fundamentally intangible. Thus, the informationization of society can be defined as the change from an era in which tangible material goods produce values to one in which intangible information produces values."[79] In these stories, the bullet train creates information that is valuable because it can be sold, because it confers power on those who control it by helping them to achieve a particular goal, or because it adds significant value to land and other commodities.

These stories also deal primarily with a different kind of information than tended to be highlighted in scholarly studies or official government reports and plans. The information at their center is often illicit: inside information used to make illegal profits; misinformation deployed to cover unseemly truths; secret information selectively displayed for pressure, blackmail, justice, and revenge. It is monopolized by economically and politically powerful or connected individuals and carefully distributed to further their own particular interests. Or it is created and deployed by criminal elements for nefarious ends. This is a very different view of the information society than the celebratory one typical of the technocrats' utopia. Rather than increasing efficiency, productivity, and equality, the informationization of Japanese society is depicted as generating new opportunities for the powerful to oppress the weak, for the dishonest to cheat the naïve, for the haves to get more at the expense of the have-nots. It was a minority perspective, but couched in blockbuster fiction with vast circulation, it reached a broad audience, giving it a power beyond its immediate scale.

Kajiyama's police procedural places the bullet train at the center of an exploration of new methods of crime and police investigation brought about by the information age. If information had value in itself and contributed value to other things, then of course it could be worth stealing and controlling through bribery, extortion, and perhaps even murder. Both the novel and the very different film version depict such information crimes (and, in the case of the novel, efforts to solve them). By showing how exclusive access to information conveyed potential economic and political power at the expense of those without it, they demonstrated that the emerging socioeconomic structures of Japanese society tended to perpetuate existing inequalities and social stratification.

The film can thus be considered as part of a wave of experimentation with new approaches to social critique that was breaking at the time not only across Japan but around the world, in a style that has been called the

[margin note: popular art the marginalized spoke]

"cinema of actuality" or "semi-documentary": fiction films that are both
artful and political, combining the art of filmmaking with the methods of
journalism and fiction to promote goals of human rights and social justice.[80] These labels are most often applied to avant-garde films, but *Black
Super-Express* brings the style to a mass audience in a studio film by using
a compelling fictional narrative to take on challenging political themes—
inequality and capitalist exploitation in the context of high-speed economic growth—and reveal unpleasant truths about Japanese society. Placing the film within a more particular context, it was also the last in Daiei
studio's "Black Series" of thrillers produced between 1962 and 1964, several
of which tell stories of industrial espionage and similar crimes. One other
film in the series, *Black Test Car* (*Kuro no tesuto kā*) not only had the same
director (Masumura Yasuzō) and leading actor (Tamiya Jirō) but was also
based on a Kajiyama novel.[81] *Black Super-Express* and the novel on which it
was based are thus part of a multilayered trend of critiquing the dark side
of high-speed economic growth through popular fiction.

Kajiyama wrote this story as a police procedural in order to explore
what he saw as a newly prominent type of crime centered on exchanges of
information and, more specifically, the new methods adopted by the police to combat and uncover such crimes. He was fascinated with methods
of investigation particular to information crimes, in which, unlike crimes
of violence, there are no bodies or weapons, and sometimes even the victim
does not see that there has been a crime. How, he asked, does such an invisible crime come to the surface? Kajiyama called this kind of intellectual
crime—in which information serves as either the mechanism or the pay-
off—a "modern weapon and a blind spot in today's world." He wanted to use
his novel about an information crime to illuminate this little-known world,
not primarily to expose the crime but rather to depict "the difficult investigations and the ingenious process" of the police agency's under-appreciated
"intellectual crimes division."[82] Several pages of the novel are devoted to an
explanation of the structure, characteristics, goals, and methods of that division. Drawing a distinction between detectives in the violent crimes division, who look for material clues like a murder weapon or fingerprints, and
those who investigate information crimes, the narrator explains that the
latter must search for these "hidden crimes" by "getting hints from rumors,
secret-telling letters, and such."[83] Investigators of information crimes must

pay attention to all kinds of information—rumors, secrets, accusations order even to discover that a crime has been committed.

Of course, the idea of using information (or misinformation) for p was not entirely new. Since the development of capitalism, access to information about a potential investment could be the difference between riches and ruin. And since the establishment of a patent system, unscrupulous inventors have tried to steal ideas. In the context of Japan's information society of the 1960s, intellectuals and policy makers were more acutely aware of information systems, and information moved faster, was vastly more abundant, and was beginning to dominate the entire national economy but continued to be largely monopolized and controlled by those in power. The bullet train was interpreted as part of the information society as much by critics of the emerging social structure as by its architects. Like urban planners who sought to use control of information flows as a tool to fix social problems, Kajiyama and other writers of bullet train fiction saw the power of information. But they highlighted the ways in which that tool was used as a weapon. While both groups sought to improve society, fiction writers did so by critiquing the consequences of unequal access to information within an increasingly information-centered socioeconomic system.

Kajiyama's novel and its film adaptation both tell a sordid story of blackmail, extortion, and murder in an illegal scheme to profit from JNR's acquisition of land for the bullet train, in which information is the coin of exchange. Inside information, misinformation, and secrets are used in various ways to gain profits and to force, cajole, or trick people into acting against their own interests or principles. The novel follows three separate but intertwined quests for information: the crime itself, a journalist's search for a story, and a police investigation. The crime begins with a complex plot by former JNR employee Nakae Yukichi. Nakae uses dirt he has dug up on a beautiful young female employee of the public corporation in charge of planning the new line (called the Shinkansen Corporation) to convince her to lure its married director into an affair. He then uses evidence of the affair as blackmail in order to obtain more information: the planned locations of new bullet train stations in Osaka and Yokohama.

That initial stage of the crime takes place before the start of the novel, which begins in media res with the second stage, in which the villain deploys misinformation about inside information in order to obtain land to

be used for the bullet train. Nakae approaches a struggling real estate broker in a quiet suburb of Yokohama about a suspicious deal to buy up a thin
strip of land. He convinces the broker to go along with the deal by making the false claim that he used to work for the construction minister and
knows through this connection that a factory will be built there. Through
the broker's mediation, he is able to buy up most of the relatively worthless land around the planned new bullet train station for what sellers feel
is a generous sum. But he soon sells the same land to the Shinkansen Corporation "for a price that would make your eyeballs pop out."[84] Though the
novel begins with the land purchase, it is focused primarily on two separate
but overlapping investigations: a freelance reporter pursues a story about
a woman's mysterious disappearance, while detectives in the intellectual
crimes division slowly unravel the massive corruption scheme connected to
Nakae's land deals. The film, in contrast, has no reporter, and the police do
not even make an appearance until the last few minutes. Instead, it explores
the many layers of the interconnected set of crimes and the multiple uses of
information toward illicit ends.

In both narratives, information is central to the crime, but it is also
the key to justice. In the novel, just as the two separate investigations—by
the detectives and the reporter—finally converge to trap Nakae, he slips
through their grasp at the airport, leaving Japan on an international flight;
in order to prevent Nakae from testifying against him, the construction
minister (who would have been implicated in the bribery and land purchasing scheme) has pressured the Foreign Ministry to facilitate his fleeing the
country, so he can wait out the remaining year until the statute of limitations on bribery expires. The novel ends with the detectives at the airport,
bitterly disappointed at having missed their chance to arrest Nakae, the
only witness against corruption in the highest level of the government bureaucracy. Spotting the reporter, the primary investigator hurries over and
beseeches him to write a thorough story about the woman's disappearance,
a central element in this complex corruption case, promising to contribute
materials he has collected through the investigation. When the reporter expresses surprise that the police are going to share materials, the detective
declares that he is handing over the materials not as a representative of the
police but "as an individual. No, perhaps as a citizen (kokumin)."[85] He sees
it as his civic duty, when corruption evades the criminal justice system, to

catalyze an alternative path to justice through the public dissemination of information. Ultimately, the only way to obtain justice for this information crime is through an information solution.

The film presents a significantly different story of information used as a tool of revenge by one information criminal against another. After losing his earnings from the real estate deal in an investment based on Nakae's lie about a factory, the film's antihero (the real estate broker, named Kikyō) tries to play the villain's own game against him, partnering with another of Nakae's victim-accomplices (the woman who seduced the Shinkansen Corporation director) to conduct a private investigation of his crime in order to get enough information to bribe him for a greater share of the profits from the original scheme. While he comes very close to achieving this end, his effort ultimately fails disastrously, resulting in the murder of his accomplice. But the information he had gathered in his quest for revenge, which he finally shares with the police, ultimately leads to Nakae's arrest. In a somewhat more optimistic ending than the novel, the film shows Nakae promising to testify against the corrupt officials. Justice for bullet train corruption, in this version of the story, will come through the courts, rather than the media.

Both endings suggest that, one way or another, the most crucial piece of information—about corruption at the highest levels of the government—will be made widely accessible. Though this can be seen, in some ways, as a happy ending afforded by the dissemination of information, in portraying a dark dynamic of information surrounding the bullet train, the novel and film provide a stark contrast to planners' bright, confident portrayals of efficient paths of communication solving the pressing problems of Japan's high-growth economy. In that way, they explicitly reject the standard celebrations of the new line and instead proffer a critique of a society focused on economic advantage at the expense of general social welfare. Rather than a building block of a more just and equitable future, the train becomes a marker of the persistence of the dark, venal social structures behind the façade of utopian rhetoric.

This story is all the more powerful (and dark) because it is not entirely fictional. In the afterword to his book, Kajiyama explains that he had been inspired by an actual case of corruption involving land acquisition for bullet train construction. Uneven access to information played a key role even in the process of Kajiyama's treatment of the crimes that it engendered. In a sense, he himself benefited from a kind of insider information, having by his own account "secretly caught" word of a criminal investigation in February 1962, well before news of the case, involving land sales around the New Yokohama Station, appeared in the newspapers. His subsequent research on the case eventually became the basis for this novel, published the following year, by which time the scandal was being heavily reported in the news media. Perhaps seeking to avoid accusations of libel, Kajiyama emphasizes in his afterword that although the novel had the feel of a roman à clef, it was pure fiction, simply motivated and inspired by the injustice he saw in real events.[86]

Contrary to that legal dodge, not only the basic outlines but also many of the details of the story hew closely to the facts of the case (though there is no hint of a murder in the real-life version). Based on leaked information about the location of the new bullet train stations, a former JNR employee engaged a local real estate broker to assist him in buying a long, narrow strip of farmland outside Yokohama for slightly more than the market price, purportedly for an automobile factory. He did the same in the area of the new station in Osaka. Investigators believed that he was able to get hold of this insider information thanks to his ties to JNR. As one newspaper report put it, "Information flows around the 'JNR household.'"[87] A month later, he sold the land to JNR for what was, indeed, an eye-popping sum of over 231 million yen, about ten times what he had paid, raking in hundreds of millions of yen in profit. Even the dramatic conclusion, with the villain Nakae Yukichi of East Asia Development Corporation slipping out of investigators' grasp by escaping Japan from Haneda Airport, echoes news reports that the case fell apart when it was discovered that the real-world perpetrator, Nakachi Shingō of Japan Development Corporation, had departed from Haneda on a U.S.-bound flight, then continued on to an unknown destination. The closing line of one of the final newspaper articles on the case, a quote from one of the detectives, is reminiscent of the novel's conclusion: "This is certainly a problem for our investigation. But we will not let this incident be swept under the rug; we will pursue it to the end."[88] Reading this, one cannot help but think of the fictional detective's determination to seek justice by sharing everything he knew with a reporter, just as the investigator quoted here had clearly done. In fiction as in life, Kajiyama

FIGURE 3. Scene of the CTC control room at Tokyo Station from *Shinkansen Explosion* (*Shinkansen daibakuha*, 1975).

seems to suggest, publicizing information can bring an element of justice for information crimes. By writing this novel, he himself was helping to make sure that happened.

Kajiyama's novel and the subsequent film both came out in the early years of the information society boom. Another film about the bullet train was made a decade later, when Japanese society was well into the information age and in the midst of a process of computerization, including the introduction of the COMTRAC system on the bullet train in 1972 with the partial opening of the San'yō Shinkansen between Osaka and Okayama and a subsequent upgrade of the system accompanying the 1975 opening of the remaining section of the line to Hakata. The operation of this computerized control system is at the center of *Shinkansen Explosion*, a 1975 action film about a train using that brand-new section of high-speed rail, the Hikari number 109 bound for Hakata, which is carrying a hidden bomb set to go off as soon as the train's speed falls below 80 kilometers per hour.[89] The film's action centers on the complex role of communications in the operation of the train, with diegetic explanations of centralized electronic information collection and control and long, slow pans of the busy control room at the Tokyo headquarters, a futuristic space full of communications terminals and blinking display boards showing the progress of every train along the entire system (fig. 3).

The constant communication of information between the control room and every train on the line features as both the system's savior and its Achilles' heel. The COMTRAC system communicates a steady stream of details about each train's location, speed, and condition to the Tokyo head-quarters. In the film, the engineers in the control room use that informa-tion to keep the path clear for the threatened train so that it can avoid slow-ing to a speed that would trigger the bomb. The directions communicated to drivers of the Hikari 109 and all other trains operating at the time re-peatedly avert disaster, but at the same time, every possible solution that the JNR employees imagine comes up against the automatic controls that, in normal conditions, permit safe high-speed operations. Each idea is re-jected when operators point out that unusual activities like opening a door or breaking connections between cars would activate the automatic brakes, bringing the train's speed below the trigger point.

Though the bomb threat in this film is potentially much more violent than the bribery, extortion, and shady real estate deals treated in *Black Super-Express*, at some level, it is also essentially an information crime. The mastermind behind the plot was the owner of a small electronics fac-tory that had recently gone under. Angry and frustrated by the economic forces that caused his misfortunes, he finds a way to survive in the informa-tion economy, albeit through a horrible crime. By planting an elaborately constructed bomb on the Hikari 109, he creates a situation in which infor-mation that he controls—the location of the bomb and how to defuse it—becomes extremely valuable. He proceeds to sell this vital piece of infor-mation to JNR for the hefty sum of five million U.S. dollars. While every other important information exchange in the film takes place through tele-communications technologies (telephones, two-way radios, facsimile ma-chines, and computerized communications about train locations, as well as the mass media of television and radio), the information purchased by JNR is delivered in an old-fashioned way: on paper, in a large envelope full of di-agrams and instructions. In a far-fetched plot point, this information is lost in a random restaurant fire, a strong statement about the dwindling feasi-bility of such old methods in the new economy.

Of course, the real crisis propelling the plot is the threat to the hundreds of lives on board the train, but this is just the motivation for the activity that the film emphasizes throughout—the delivery, redirection, and withholding of information. The symbolism of information flows and blockages suggests that the bullet train, too, might become a relic in the evolving information society. Electronic communication between the headquarters and the trains under its control is always clear and immediate, enabling the split-second timing needed to avoid crashes while keeping the bomb from exploding. In contrast, the passengers on Hikari 109—the carriers of information in both Umesao's theory and Tange's megalopolitan vision—experience communication blockages, bottlenecks, and slowdowns. The first obstacle to information is raised by the JNR management dealing with the disaster from Tokyo. As part of their strategy for handling the threat, they initially try to contain information about the reasons behind the train's slow speed, so as not to cause a panic among passengers. In spite of such obfuscation, the conductors' repeated searches for the bomb hint to the passengers that there is a serious problem. For many passengers, the slowdown is felt in terms of their own function in transmitting information, as electrons running along the Tōkaidō spinal cord. As soon as it becomes clear that there is a problem, passengers crowd the on-board telephones. This communications technology, which had been a bragging point of Tōkaidō Line expresses even before the bullet train, became yet another information bottleneck, as passengers desperate to communicate to others outside the train try to elbow each other out of the way or pay to jump the line. Later, a businessman on the train becomes dangerously obsessed with getting the important information in his briefcase to his business in Osaka. While the film highlights the effectiveness of the computerized communication system underlying the operation of the bullet train system as a whole, it presents a potential failure of the train itself as a communications infrastructure. Though the information age transformed the train from an industrial tool to an information system, the continued development of telecommunications technology undercut its importance in the information society.

While Umesao Tadao is often credited with originating the idea that Japan was becoming an information industry society, the changes he identified were inescapable, and many people noticed and incorporated them into their thinking about Japan's present and future. Trains were a key infrastructure for the development of industrial capitalism in Japan in the late nineteenth and early twentieth centuries, and looking back from the twenty-first, the tendency is to see information systems in terms of computer networks and digital communications. However, in the transitional moment of the 1960s, when telecommunications technologies were rudimentary, but futurologists glimpsed the dawning of the information age, the unprecedented speed of the bullet train brought the railway into the discussion. As the information age began to emerge from the long shadow of the industrial age, trains, as symbols of speed and connection, were open for appropriation by those who sought to articulate and imagine the new era. As familiar tools that could be repurposed for changing times, the language and rhetoric around industrialization was useful for envisioning what the new information society might look like.

As the discourse on information society began to attract mass attention in the 1960s, national, regional, and urban space came to be viewed in terms of networks of connection, a concept readily applicable to railways, which themselves form a physical network. Thinking about the bullet train in terms of information helps explain why, in the "jet age" of the 1950s and 1960s, the Japanese government would make a massive investment in what many were calling a sunset industry. Japan's transformation into an information society made planners re-envision rail from a carrier of goods to a carrier of information. This updated conception put high-speed rail on the cutting edge, along with other communications technologies. While politicians and planners took advantage of this rhetoric to promote their own ideas, their critics viewed the bullet train from a different angle in order to present alternative views of contemporary Japanese society and its potential futures.

An important ingredient of imagining the future, paradoxically, can be rethinking the past. For that reason, the process of envisioning the future created by high-speed rail spurred the retelling of history. New narratives reimagined the past as a precursor to a particular understanding of the present and vision of the future. The next chapter examines the ways in which, just as the bullet train informed futurology, it also prompted a gaze backward in time: Japan's railway successes of the 1960s brought back memories of the role of super-express trains in war and empire, complicating efforts to come to terms with that history. The bullet train thus played a role in creating a nostalgia for Japanese empire while reinforcing certain ideas about its potential role in Asia.

Edited/selective history

4   Nostalgia for Imperial Japan

WHILE THE NEW TŌKAIDŌ LINE informed competing visions of Japan's future, its shadow also extended back in time, inspiring a reimagination of wartime experiences. Public discourse around the new line connected it to two railway systems of the past. When the South Manchuria Railway (SMR) Company's "Asia Express" started running in 1934, it was the fastest in Asia and more luxurious than any in Japan. Five years later, anticipating that the expansion of the empire and escalation of the war in China would increase demand for ground transportation at home, the Japanese government approved the construction of a new high-speed trunk line between Tokyo and Shimonoseki. Popularly dubbed the "bullet train" (*dangan ressha*), it was to run even faster than the celebrated Manchurian train, pulled by a more powerful locomotive across more technically challenging terrain. By the end of 1943, however, the war's insatiable appetite for human and material resources had brought to a halt both construction of the bullet train line in Japan and operations of the luxury express train in Manchuria. Both largely disappeared from the public imagination until the late 1950s, when plans for a new railroad system that promised to be the world's fastest (and revived the "bullet train" nickname) prompted a wave of recollections of that earlier moment of Japanese railway prowess.

Juxtaposition of wartime descriptions and postwar memories of these two railway systems reveals the expansiveness, longevity, and durability of the aesthetic function of infrastructure, which can extend beyond its technical operation in both time and space.[1] The aesthetics of the Asia Express stretched out spatially—railways in Japan suddenly seemed shabby and slow by comparison—and anticipation and nostalgia extended the affective impact of the wartime bullet train chronologically in both directions. Expectations about its significance formed before construction even began, and memories of both wartime trains colored perceptions of the postwar New Tōkaidō Line as it was being planned and built many years later. Pride in the engineering achievements that went into the Asia Express lasted beyond its final run in 1943 and were brought to the surface alongside complicated emotions surrounding the unfinished bullet train by the 1958 decision to build a standard-gauge high-speed railway along the Tōkaidō route. In these ways, the periphery helped shape the metropole, and nostalgia for the imperial past imbued anticipation of the future. In both historical moments, infrastructure represented national progress. Railway projects in Manchuria and Japan helped reinvent the national spaces they (potentially) traversed and connected, accruing new meanings in connection with particular ideals for the future, meanings that later colored memory of those projects and their connection to the present.

Considering the war years alongside the 1960s also shows how high-speed rail was incorporated into a wide-ranging reimagination of wartime history and its relation to the present that was taking place at the time, a dynamic that boosted excitement about the bullet train. Paul Noguchi argues that the railroad is a particularly strong site of nostalgia in Japan, perhaps due to its importance in everyday lives.[2] In the 1960s, rail-centered longing for a lost time and place merged with an increasingly common approach to retelling the past. As Yoshikuni Igarashi demonstrates, postwar Japanese society grappled with the trauma of loss and radical change by creating narratives that transcended the rupture of defeat and occupation.[3] By telling stories that explicitly or implicitly connected the Asia Express and the wartime bullet train plan to JNR's present-day high-speed rail project, writers helped to knit back together the wartime and postwar periods that had been sliced apart by the drastic changes in Japanese society since 1945 to tell one continuous story of technological and infrastructural development.

Of course, this is just one aspect of a wide range of uses of public memory toward varied ends in the early postwar decades. Some used attention to Japan's autocratic past or a focus on personal war responsibility to oppose government actions and policies, such as activists protesting the revision of the U.S.-Japan Security Treaty in 1960.[4] In contrast, public memories of the Asia Express, produced mainly by former SMR employees or others directly associated with the train, tended to shine a favorable light on their subject. Celebratory stories of the Asia Express, silent about its role in promoting Japanese control of Manchuria, in some ways absolved their narrators of guilt for their participation in Japanese imperialism. Memories of the original bullet train similarly focused on the suffering of Japanese people on the home front while ignoring the realities of aggression and domination. By stripping these events of their violent contexts, such memories rested on and perpetuated a narrow view of war responsibility that depicted the Japanese people as the victims of militarist leaders, who bore full blame.[5] Memories of imperial and wartime express trains that left out empire and war thus helped to foreclose the possibility of real justice by contributing to the resilience of the system structured by postwar adjudication of war crimes and Cold War rehabilitation of Japan.[6]

Whether rosy or painful, such recollections are both individually inflected and oriented toward the present; they are created in a specific historical moment by individuals holding a particular social position in both the moment remembered and the moment of remembering.[7] Various actors with differing agendas, each telling their own narratives of the past, collectively create a kind of truth, "a commonsense view of the past and its meaning."[8] Their stories reflect not only personal experiences and interests but also the circumstances of their reproduction: the wartime memories sparked by the bullet train, therefore, were "deeply embedded in the material conditions of postwar Japan."[9] Several layers of context shaped these railway reminiscences. Most directly relevant was the widespread excitement about infrastructure inspired by the bullet train as well as the new monorail and expressways under construction. More broadly, rapid economic growth and technological prowess were contributing to Japan's rising international status, and the Japanese government was building a new position in Asia as a leading provider of technical aid and expertise. With these changes came a re-evaluation of the war and competing memories of the aims and impact of colonialism. The socioeconomic and diplomatic changes of the 1960s contributed to resurgent nationalism and increasingly positive portrayals of the war in both scholarly histories and public memory.[10]

Narratives of the past were shaped to some degree by the genres through which they were told and the audiences for whom they were produced. Stories of wartime trains could be found in both fictional and nonfictional forms on national television, in popular history books, and in specialized publications for railway enthusiasts and professionals. For each particular example, the interests and expectations of the intended audience helped determine both content and form. For instance, while an official record of the construction process aimed at railway professionals was necessarily detailed and factual, one geared toward a popular audience could take a more broad, sweeping view. Material produced for railway enthusiasts focused on the specifications of cars, tracks, and operations, while a television drama took artistic license in order to create a compelling story. For these reasons, fictional narratives had a potentially broader audience and stronger emotional impact, but all contributed in different ways to a shared memory of the past, re-interpreted in terms of their own present.

Postwar stories about these wartime trains worked against the possibility of Japanese society coming to terms with the violence and oppression of its imperialist past. They did so in two ways: by editing out unflattering parts of the story and by focusing in closely on the "inner territory" of the Japanese empire. The first approach tended to elide the fact of violence against others through soft-focus, carefully cropped images of the past. This kind of selective memory, involving a constant process of forgetting, carefully trimmed out unpleasant elements to "render memories of war into benign, nostalgic form."[11] Remembrances of the Asia Express, for instance, brought forward a moment of success for SMR from an exclusively Japanese perspective. Forgoing the opportunity offered by the passage of time to add a broader or more nuanced analysis, they mostly echoed the excitement and praise that marked 1930s promotions and representations of the train without considering the real reasons for or consequences of the Japanese presence in Manchuria. The second category of reminiscence put the spotlight on the sacrifices and hardships of the war years, but again with a carefully circumscribed view, encompassing only the Japanese people on the home islands. Stories about the bullet train presented the accomplishments of Japanese railway planners and builders, focusing on the obstacles, challenges, and hardships they faced in order to highlight their national spirit and persistence against adversity. Such tales of "achievement against the odds" worked to heighten the "enchantment" exerted on the national imagination by high-speed rail.[12] These stories depict the pain of war, but the only victims appearing within the frame are Japanese.[13] Violence inflicted upon others by Japanese military forces or colonial apparatus is entirely absent.

In both cases, postwar writers were not creating completely new lenses through which to view these trains or providing entirely novel interpretations. Rather, they reproduced the rhetoric that surrounded the projects in their own time, a rhetoric that was saturated with the ideologies supporting Japanese empire and aggression. Writers were certainly recalling their own experiences (or, in the case of fictionalized stories, imagining the historical lived experience of others), but they expressed those memories very much in the terms in which those events and achievements had originally been interpreted and presented. This repetition brought reverberations of imperialist rhetoric into the postwar world. Nostalgia for wartime railways to some extent rewrote the past, but it also dragged elements of the past into the present, translating them into a new historical context.

During the war years, the representation of these railway systems was shaped by what Aaron S. Moore identifies as the "technological imaginary" that Japanese leaders developed by blending "creative imagination and technical expertise in the formulation of wartime and colonial policies, as well as the construction of numerous large-scale infrastructure projects designed to incorporate the hopes and desires of various peoples."[14] Janis Mimura argues that these leaders believed that Japan could bring about co-prosperity and overcome resource deficiencies through superior technology and national spirit. And fast trains were an important example of the modern technology that they thought made Japan the natural leader of Asia.[15] Such ideas were dredged up again by postwar nostalgia. The profound changes in Japanese society in the decade after defeat created a sense of a new beginning, and claims that a "new Japan" was emerging were ubiquitous in the 1960s. But at the same time the economic and industrial successes of the day provided material for stories of technological progress that recovered connections with the wartime past. The powerful Asia Express was prominent in imperialist propaganda as a symbol of Manchuria's flourishing industry under Japanese management. And the construction challenges of the wartime bullet train were mobilized to show both technological superiority and a national spirit that could overcome any obstacle. Nostalgic representations of both trains gave those wartime ideas new life in the context of the peace, democracy, and prosperity of 1960s Japan.

RIDING THE "ASIA": POSTER TRAIN OF THE SOUTH MANCHURIA RAILWAY COMPANY

The Asia Express was an impressive sight, its powerful, streamlined locomotive pulling elegant cars at surprising speed across the plains of Manchuria. Carrying passengers back and forth between the southern port of Dalian (called "Dairen" in Japanese and most English-language publications during the period of Japanese control), the puppet-state capital Xinjing, and the northern city of Harbin, the train simultaneously performed a separate and very different task as a tool of empire. Promotional materials consistently reminded readers at the time that it was the fastest train in Asia and on par with the speediest in Europe and the United States. This status, along with its cutting-edge technologies and luxurious amenities, made it a prime example used by Japanese colonial administrators to draw a picture of successful development in the region. Its demise in 1943 drew little attention, and the train was largely forgotten by the Japanese public for the next twenty years. But when JNR made a plan to build the world's fastest train in Japan, national pride in achieving a railroading milestone catalyzed an outpouring of reminiscences about the earlier feat of Japanese engineering in Manchuria. In both discourses—contemporary celebrations of the Asia Express and postwar recollections—a narrow focus on the machinery of the train itself obscured its imperialist underpinnings. The people who were directly involved in building and operating the Asia Express (and perhaps, therefore, imbibed the propaganda surrounding it most deeply) later provided the main voices in the public recollections of the train sparked by the construction of the bullet train. An examination of the words and images used to promote the Manchurian train as a success story of Japanese control reveals the rhetoric of imperialism that would become the foundation of postwar nostalgia for the railways of empire.

The Asia Express was a star performer in Japan's imperialist project in Manchuria, which rested in part on claims that Japanese control benefited

Manchuria through development of technology and industry. Yoshihisa Tak Matsusaka has categorized the energetic expansion of the entire rail system by the SMR as a version of the "railway imperialism" commonly exercised by the Western powers in the late nineteenth and early twentieth centuries.[16] Japanese leaders used infrastructure and engineering expertise not only to strengthen their physical control of Manchuria but also to project images of success there to audiences in Manchuria, Japan, and Western nations. The steam locomotive had been a symbol of modern transformation in Japan from the late nineteenth century, and it similarly came to represent Japan's imperial mission in Manchuria in the twentieth.[17] Matsusaka argues that "conspicuous displays of engineering skill" in SMR's railway and industrial facilities furthered the state's "civilizing mission" and presented a positive view of Japanese control to both foreign and domestic observers. The SMR sought to cultivate this image through promotional materials depicting "late-model locomotives hauling sleek trains across the Manchurian plain, steel trestles spanning the Yalu River, and the towering smokestacks of the Anshan Ironworks."[18] Historians have called the Asia Express SMR's "poster train" or "company mascot," featured in pamphlets, posters, postcards, photo collections, guidebooks, travel literature, and even a postage stamp.[19] Louise Young considers its significance for the imperialist project in Manchuria: "Surpassing records in Japan and matching the railway technology of the West, the Asia Express became the symbol of an ultramodern empire where technological feats opened up new vistas of possibility for Japan."[20]

These scholars point toward a dynamic that might be explained in terms of the aesthetic function of infrastructure. The Asia Express was designed and built to move people rapidly from one city to another, but it also performed a kind of social display, conveying meanings and evoking emotions. So the train was both a symbol and a tool of empire, representing the accomplishments of the Japanese-controlled state of Manchukuo and at the same time contributing to its stability and control by bolstering support and undercutting opposition. The combination of technical achievements and aesthetics of the Asia Express made it powerful as a subject of propaganda in the 1930s and as an object of nostalgia in postwar Japan. The quantifiable and comparable measure of average and maximum speeds put the train squarely in the category of global leaders, while its streamlined

FIGURE 4. Posters for the Asia Express. The poster on the right reads: "Stream-lined special express Asia: Dalian to Harbin, a rapid 13 hours." (Minami Manshū Tetsudō Kabushiki Gaisha, *Ryūsenkei Tokkyū Ajia* and "Minami Manshū Tetsudō," *13-ban madoguchi*.)

locomotive and cars, the latest fashion in railway design at the time, gave it a cutting-edge, modern look. In addition, the technical problem solving that reduced noise and vibrations in order to provide a comfortable ride even at high speeds and in varied weather conditions, including climate control in all cars, constituted a bragging point that promoters never failed to mention. And while these technical aspects were deployed to show off the superior skills of Japanese engineers, the train's elegant style and luxurious amenities contributed to the feeling of traveling in comfort and style, aspects that also formed a key part of its international reputation. The precise punctuality of its operations—a former crew member boasted that one could set a clock by its arrival—showed off the SMR's effective management as well as the system's efficient technical operation.[21]

Whether in English or Japanese, descriptions of the Asia Express that appeared in publications produced in Japan, Manchuria, and the United States all include more or less the same basic elements. The posters shown here highlight two of the most prevalent: the train's speed and its modern, streamlined form (fig. 4). The locomotive is central in both posters, which emphasize the streamlined shape both visually and in writing, while the lettering evokes modernity (through a stylish font in English) and speed (through the addition of speed lines to the Japanese characters to suggest they are racing past along with the train itself). Written descriptions invariably told readers that the new train was number one in Asia and on a level with the best trains in the United States and Europe in terms of speed, size, and amenities. Though speed was the most easily and precisely measurable element of its quality, aesthetic design features were also important: with its streamlined form and luxurious appointments, it was constantly referred to in Japanese descriptions as "gallant" (sassō) and in English-language descriptions as powerful, modern, and even beautiful.[22] Technological advances that warranted mention included features essential to high-speed operation, such as the automatic stoker (which could feed coal at a fast enough rate to maintain the necessary steam levels) and the use of special lightweight metals.[23] But even more excitement and adulation surrounded the technologies contributing to a comfortable ride, most notably the air-conditioning system and measures to reduce vibration and noise inside the passenger cars.[24]

Japanese descriptions noted the Manchurian sourcing of materials for the train and the Japanese engineering and design expertise behind it.[25] This reflected the typical division of labor in an empire: expertise from the metropole, raw materials from the colonized periphery. Many accounts highlighted the hard work and efficiency of SMR employees, especially in reference to the short time it took for the design, construction, and testing of the train—less than a year from conception to implementation— which they attributed to the dedication, passion, and cooperation among the SMR's design and construction teams. For instance, a writer for a popular Japanese science magazine called the Asia Express a "home-made train, designed and constructed ourselves."[26] Though the train was physically located in Manchuria, SMR's Japanese management and workforce muddled its national identity. Nevertheless, the achievement of SMR's engineers in designing and building one of the fastest trains in the world was clearly a point of pride, as evidence of Japan's success in the management of Manchurian development and the high level of its technological capabilities. Two of Japan's biggest daily newspapers, the Asahi shinbun and the Yomiuri shinbun, placed the train in the context of the global competition for railway speed. The Asahi reported the train's inaugural run as a "brilliant start for the world's third-fastest train," and both papers looked to a higher global rank in the future, calling the Asia Express a step toward the "number one fastest train in the world," its speed even rivaling that of standard passenger planes.[27] Beyond such news of opening day, the *Asahi* maintained popular awareness of the train among Japanese readers through reports of new developments and occasional articles touting the pleasures of a journey making use of this fast, comfortable, and well-appointed train.[28]

Descriptions of the Asia Express aimed at Japanese audiences tended to promote Japan's role in Manchuria not only by demonstrating successful development but also by fostering an emotional connection to the physical territory among Japanese readers. In her study of tourism in the Japanese empire, Kate McDonald shows how the infrastructures of tourism fostered popular Japanese support for the nation's role in Manchuria, as itineraries and guidebooks created a social imaginary that included both Manchuria and Korea within the nation by influencing what travelers saw and how they understood it. Introducing tourists to famous battle sites of the Russo-Japanese War, for instance, fostered an emotional connection to Manchuria, while pointing out new factories, bustling modern cities, and productive fields highlighted the benefits of Japanese control.[29] Stories about riding on the Asia Express—which typically described the same mix of important battle sites and scenes of industrial and urban development, agricultural cultivation and experimentation, and vast, empty plains through which visitors from Japan were typically guided by tour books and travel itineraries—could serve as a substitute for tourism, allowing readers to experience travel through Manchuria vicariously and broadening the audience for such pro-colonial displays.

Some level of familiarity with the distant train permeated the entire nation when a chapter titled "Riding the 'Asia'" was included in the 1937 edition of the Ministry of Education's fifth-grade Japanese reader.[30] With a four-page description of a child's first ride on the famous train in the national textbook, used in every elementary school in Japan, children in all corners of the nation could imagine themselves on a high-speed train ride from Dalian all the way to the northern terminal in Harbin. This meant that an actual journey to Japanese-controlled Manchuria was not an essential ingredient of either a feeling of connection at that time or postwar nostalgia for the Asia Express. The textbook brought the train into the lives of all elementary school students.[31]

The textbook's first-person narrative of a young Japanese boy traveling on his own from his home in Dalian to visit his uncle in Harbin begins by identifying the train itself as an object of desire. The boy narrating the journey declares to his readers that he had long been hoping for an opportunity to ride the "Asia." Through the description of his journey, the narrator introduces the defining characteristics of the train, echoing the many newspaper and magazine reports about it. Seated on the train, waving to his mother on the platform, the boy can see that she is saying something to him, but inside the insulated car with its double-paned windows, he cannot hear her voice; the noise of the platform only briefly intrudes on the quiet interior of the car when the door opens for another passenger to enter. The train departs precisely at the scheduled time of 9:00 a.m., pulling away from the platform so smoothly that it seems to be floating. It gets colder outside during the high-speed northward journey, and the boy overhears the conductor calling for an increase in the temperature, a nod to the train's famed heating and cooling system that provided for year-round comfort. The story ends, of course, with the boy's safe, to-the-minute on-time arrival in Harbin, where he is greeted by his uncle's proud and happy "You got here all on your own!" as he exits onto the moonlit platform and into the cold night air of northern Manchuria.[32]

Along the way, the conductor, fellow passengers, and the narrator himself call attention to important landmarks the train passes, the same sights noted in any contemporary travelogue or guide. Cultivated fields, agricultural experiment stations, industrial sites, and impressive cities demonstrate successful development, while the wide-open plains connote a region ripe for immigration. Like many other descriptions of sightseeing through the train's windows, the textbook highlights a stone monument marking the battlefield death of Lieutenant Nogi Katsusuke (the eldest son of General Nogi Maresuke) during the Russo-Japanese War, a battlefield death that inspired a well-known poem by his father.[33] As McDonald argues, tourism to such battle memorials linked Japanese travelers affectively to the space of Manchuria as territory where Japanese blood had been spilled.[34] It is likely that the virtual visit of schoolchildren through the story in their textbook had a similar impact.

Images of the SMR's poster train also appeared in propaganda aimed at building support for (or at least reducing opposition to) Japanese imperialism among a Western audience by suggesting that Japanese control contributed to the development of Manchuria, claims that were then repeated by both Japanese advocates and sympathetic non-Japanese writers and publishers. A ride on this train was an essential part of any visit by a foreign dignitary or reporter, its importance reflected by the fact that visitors' impressions of the train became news. The Dalian-based English-language paper *Manchuria Daily News* reported comments on the new train by, for example, a Shanghai-based American journalist, a French railway expert, and the leader of an Italian Fascist goodwill mission, the last of whom used the Asia Express as a symbol of Japanese-controlled Manchukuo to argue in favor of the technocratic state.[35] Giacomo Paulucci di Calboli Barone Russo, who had served in the League of Nations Secretariat before moving to the propaganda body of the Italian Fascist regime, stated that "this wonderful train represents in my opinion the symbol of the progress realized already throughout" Manchukuo. He then continued in a more general vein, comparing the modern state to a locomotive: powerful, but with complex and delicate workings and, therefore, requiring competent men to run it with iron discipline along a well-planned path. He concluded that just "as the Asia Express gets smoothly and punctually to all appointed places, in the same way your country will reach its goal safely and rapidly."[36] This fast, powerful train served well as the representative of Manchuria's economic development, but also as a metaphor for the puppet state.

The most talked-about group of visitors to ride the Asia Express, both at the time and in postwar remembrances, was a delegation of American newspaper reporters who visited Manchuria in the fall of 1934 and were invited on board for a test run in the weeks before the new train began regular operations.[37] Given the international condemnation of Japan's invasion of Manchuria and creation of Manchukuo (the impetus for Japan's withdrawal from the League of Nations), one might expect developments there to be viewed with a disapproving eye in the United States, but American reporting on the Asia Express contained only a subtle hint of criticism, if any. The test ride resulted (at least) in articles in the *New York Times* and *Chicago Daily Tribune*, both of which sounded more or less like SMR press releases. Both newspapers called the Asia Express one of the "crack trains" of the world and noted its streamlined form and its status as the fastest in Asia. The *Tribune* reported that it was built entirely in Manchuria by SMR's own shops, and the *Times* reporters "noticed that designers of the train had been successful in reducing vibration to a minimum."[38] Such reports suggest that efforts to have the train star in cultural diplomacy initiatives that would build respect for the puppet state and support for the SMR's role in Manchuria enjoyed a certain amount of success.

The train also made an appearance the following year on a new American newsreel series called "March of Time." An eight-minute segment about Manchuria has a schizophrenic feel, presenting Japanese control alternately as locally beneficial but globally threatening. The segment begins with a reminder of the Japanese invasion of Manchuria and subsequent withdrawal from the League of Nations in response to criticism of the establishment of Manchukuo but then turns to descriptions of a booming economy, industrial growth, and innovation in agriculture. While this suggests that Japanese control was a boon to Manchuria, the narrator warns that these advances are "making 'Manchukuo' into a base for further conquests." The introduction of the Asia Express, in particular, captures the tone: "Proud symbol of Japan's drive and progress is the Asia Express, modern and speedy, which pulls out of Dairen Station every day carrying swarms of Japanese officials and their families into the interior." Though starting on a note of praise, the image of Japanese people swarming into Manchuria has a strong undertone of "yellow peril"–style warnings of Japanese expansion. But the narration then continues again in the vein of Japanese propaganda, stating "To open up the country and lace it together, Manchukuo's railroad system is being extended at the rate of five miles a day. From squalid huddles of flimsy huts, Manchukuo's cities are being built up at high pressure with steel and concrete. The capacity of Manchukuo's iron and steel works is being doubled."[39] Though American evaluations of the Asia Express were not as breathlessly admiring and enthusiastic as Japanese reports, they nevertheless in some ways performed the anticipated task of promoting an image of effective management of Manchurian development, even if they did not necessarily work to ameliorate opposition to Japanese control.

A more subtle critique of Japanese rule comes through in a May 1936 article in the British periodical *The Spectator*, which was reprinted in full two months later in the Tokyo-based English-language newspaper *Japan Times & Mail*. The author, Ralph Morton, was a Christian missionary who had lived in Manchuria since 1926 and would soon return to Great Britain in 1937.[40] The title suggests an expansive journey: "Grand Tour in Manchukuo."[41] But the narrative never actually strays from the railway system, focusing on the Japanese traveler's practice of obtaining commemorative stamps at each of the important railway stations.[42] Initially, Morton's description of the Asia Express seems to resemble other celebratory articles, calling the stamp-collection souvenir a "testimonial to the speed with which Manchukuo can be seen" and imagining the Japanese traveler to be "impressed by the colossal achievements of his own country," of which this train is the most spectacular, streamlined, air-conditioned, "and superbly sprung." Because the quality of the train is equivalent to the best in Western countries, "the foreign traveler feels at home and blesses the railways of Manchukuo." Morton emphasizes the diplomatic function of the luxurious and powerful train, given that travelers' comfort shapes their perceptions of the country. But he asserts that foreigners are not the target audience of this effort. Instead, he sees it as directed at the Japanese people, who feel unwelcome in the West but in Manchukuo "are free to travel with all the superiority of the nineteenth-century Englishman in Italy, but with an added sense of possession and of their country's achievement."[43] Indeed, most Japanese materials conveyed an impression that the only people riding the Asia Express were Japanese residents and soldiers, along with a handful of Russian travelers and the occasional visiting foreigner.

To the Japanese traveler, says Morton, the people of Manchukuo, whom they make no effort to understand, are just "part of the picturesque background." Confining their travels to the familiar, Japanized spaces of the railway zone, they are protected from any uncomfortable challenges to their preconceptions. Playing on one of the points of pride of the Asia Express, he comments, "It is not only the air that is conditioned. We travel nowadays in a conditioned atmosphere. What are trains and steamers but parcels of conditioned thinking careering up and down the earth?" The Manchuria that remains invisible to the upper-class Japanese traveler is present on the same trains but riding in a different car: the third-class passengers, including Chinese, Russian, and Korean as well as less wealthy Japanese passengers. Though out of sight of the first-class passengers at the other end of the train, "they are there, traveling at the same speed and to the same destination." These passengers are not interested in stamping a book but are eager, instead, to get home, away from the Japanese towns surrounding the stations and beyond the railway. "But it is not to see them that the tourist travels. He descends from his carriage, advances to the table, stamps his book and goes back, virtuous and superior. He is seeing Manchukuo."[44] The *Japan Times & Mail* reprint gave the article a positive spin with an added subtitle: "Japanese Influence in the New Empire Has Made It Possible for the Tourist to See Country in Comfort."[45] But the absence of such tone-setting in the original lets the critical meaning—such as the sarcasm in the statement that a stamp-collecting train passenger is really seeing the country—shine through more clearly to its original British readership. Beginning as a lighthearted glimpse of the Japanese-controlled network of trains spreading rapidly across Manchuria, the article ends with a wry comment on Japanese colonizers and the hypocrisy of their claims to be helping Manchurians, when their real focus is on their own interests. Like the "March of Time" newsreel, Morton's article reverses the propaganda value of the Asia Express by placing it in the context of Japanese aggression and expansion on the continent.

The train's fame was enough that its distinctive engine was included in a collection of photographs of the world's top locomotives, edited by Raymond Loewy, an American famous for his design of a streamlined steam engine for the Pennsylvania Railroad. Loewy offers only mild praise for the engine, noting the "interesting front end" on a "rather nice engine" that "denotes careful study of details" but with an inadequate smoke deflecting scheme.[46] Regardless of its evaluation of the train itself, the book implicitly supports the claim that Manchukuo was an independent state by listing it separately in a book organized by nation. In this way, the locomotive's aesthetic function includes a contribution to Japanese propaganda about the independence of Manchukuo.

A different and quite narrow audience for propaganda about the Asia Express consisted of the very people most closely connected to the train in their daily lives: the SMR employees who designed, built, and operated it. Through their in-house journal, published by the Railway Office's Personnel Department, the SMR inculcated a sense of responsibility and ideals of hard work and devotion to the company. A header on the journal's table of contents page listed the "Five Precepts of the Railway": devotion and service, harmony and concentration, strict observance of regulations, study and training, and simplicity and fortitude.[47] This kind of statement of principles helped shaped railway employees' understanding of their own efforts and their role in the company's activities. An article about the Asia Express in the journal's June 1940 issue presents a litany of the standard praise for the train: its gallant, dignified appearance; the superior rolling stock and famous streamlined locomotive; its speed and safety; and the comfort provided by its advanced air-conditioning system, vibration reduction methods, soundproofing, and luxurious amenities in the observation and dining cars. It also highlights the ability of SMR engineers to solve problems through technology, as well as the global attention being paid to the train. Aimed at railway employees, however, the article gave particular attention to praise for the "selfless efforts of every regular employee" who had contributed to the train's excellent reputation and for the passion, responsibility, and company pride of crew members.[48] These concepts ultimately shaped the postwar nostalgia for the train because SMR employees were the primary "experts" called upon in the 1960s to publicly reminisce about their experiences building and operating the otherwise mostly forgotten train.

MEMORIES OF SPEED IN
THE ERA OF THE BULLET TRAIN

Postwar reminiscences of the Asia Express bear a striking resemblance to the discourse of the 1930s in tone, rhetoric, and claims.[49] This had the effect of representing the history of Japan's colonization of Manchuria in positive terms of development. Those retellings that ventured outside the train itself did so in a passive voice that painted the events of the Manchurian Incident and Japanese control as natural and inevitable. In a historical moment of resurgent nationalism, these memories of the war worked to paper over ugly truths about the past at a time when many people were eager to forget them. This perspective also resonated with the new role in Asia that the Japanese government had begun to build in the 1950s through development projects and technical aid programs.[50] The timing of this resurgence of interest in the railways of the SMR in general and the Asia Express in particular overlaps with the planning, construction, and early operation of the New Tōkaidō Line, beginning soon after the government's approval of the plan in December 1958 and lasting for just over a decade. Most reminiscences mention the bullet train, connecting it to the Asia Express through comparison and thus helping to bridge the divide of defeat.

Public remembrance of the Asia Express first appeared in publications devoted to railways but expanded out to more general audiences after the opening of the bullet train generated more popular excitement about fast trains. The trend began with a February 1959 article and photo essay in a journal for railroad enthusiasts, *Railroad Pictorial* (*Tetsudō pikutoriaru*), by Hokkaidō University professor Oguma Yoneo.[51] The same journal published an expanded special issue dedicated to the SMR in August 1964; this includes descriptions and photographs of several trains and locomotives, but the "Asia" is clearly the star as the subject of about half of the articles and a third of the photographs, much more than any other single train.[52] Three contributors to this special issue—all former SMR engineers—appeared two years later on a television show devoted to the Asia Express, part of a long-running series that highlighted personal reminiscences of important historical events of the preceding decades. The television show brought their stories to a much wider audience than the readership of a specialty journal for railway enthusiasts, and that audience was further extended by the publication of the show's transcripts as a book three years later.[53] Some of these voices were echoed yet again in a 1969 book inspired by the approaching centennial of Japan's first railroad, *One Hundred Years of Speed* (*Supīdo hyaku-nen*). Exploring a century of Japanese railway history, the book positioned the New Tōkaidō Line as the culmination of the constant improvement and acceleration of Japan's railways over this era of speed and placed the Asia Express (as well as the wartime bullet train plan) squarely in the middle of it.[54] As the Asia Express "boom" receded, the audience narrowed back to railway specialists; recollections moved from national television and general-interest publications to more niche venues, such as a 1970 collection of photographs and blueprints of SMR trains and a 1971 history of railway technology development by the SMR.[55]

People who remembered reading about the Asia Express in the 1930s might have heard a familiar ring in Oguma's February 1959 reminiscence. It starts off as a travelogue, mentioning all of the same sites that were described in wartime guides to train-window tourism and thus, potentially, reinscribing the feelings those materials originally aimed to cultivate. Its present-tense description of the journal from Dalian to Harbin, noting the details of what passengers would have experienced on the train and seen out the window, gives readers the feeling of being on the train, the same kind of vicarious experience that elementary school students might have felt as they read "Riding the 'Asia'" in their textbooks twenty years before. Like that 1930s story, Oguma's description evoked many of the train's oft-noted characteristics as if through the eyes of a passenger: the size and majestic appearance of the locomotive, the noise of the bustling platform being blocked by the double-paned windows, the smooth and quiet ride, the air conditioning that created a comfortable atmosphere on the train in any season. Among the important historical sites Oguma notes is the area of the September 18, 1931, explosion on the railroad track at Liutiaohu, which, he explains, "became the fuse for the Manchurian Incident," using the Japanese euphemism for their army's invasion of northeast China, which culminated in the establishment of Manchukuo.[56] He neglects to mention that the explosion had been set by the Japanese Guandong Army in order to create a pretext for a supposedly retaliatory attack against Chinese forces. This kind of "history in the passive voice" gives the sense that the Japanese invasion of China was something completely disconnected from the Asia Express train carrying this imaginary passenger.[57] It is as if the creation of Manchukuo was indeed something that the Japanese people watched from within a train's airtight passenger car, gazing out through the windows at events in which they had no role. In verbiage that was fairly typical of such histories, Oguma explains that Manchukuo then "appeared," having been "newly born out of the Manchurian Incident," and that it was this birth of a new "nation" that had created the need for a high-speed rail connection between the port city of Dalian, the commercial center of Mukden (present-day Shenyang), and the new capital at Changchun, renamed Xinjing ("new capital").[58] Couching an echo of 1930s writings on the train in this passive-voice explanation of its history, Oguma provided a happy memory without requiring readers to confront feelings of guilt, complicity, or responsibility for its imperialist context.

Because most of the people publicly reminiscing about the Asia Express were former SMR employees, it is not surprising to find that their memories of the train reflect both pride in their work and ideas that were inculcated during the years of its operation. Contributors to the *Railroad Pictorial* 1964 special issue and the 1966 television show quote the SMR company song in order to illustrate the feelings of what one former Asia Express driver referred to as "SMR people" (*Mantetsujin*).[59] These men highlighted the "company spirit" of SMR employees—calling it "SMR spirit" (*Mantetsu-damashii*), "the SMR man's pioneering spirit" (*Mantetsu-man no kaitaku seishin*), or "the great spirit of the establishment of the SMR" (*Mantetsu no sōgyō no dai-seishin*)—through stories of skill, cooperation, and hard work. Praise went to employees involved in all aspects of the line's success, from the designers and builders—said to have worked overtime without rest, day after day, picturing the "Asia" even in their dreams, and therefore putting their hearts and souls into creating it—to the crew who operated it and "station masters on the platform to observe the train [who] kept their eyes open even in the dust whirling around it as it passed."[60] In these ways, the recollections of former employees, focusing narrowly on their work, skirted any sense of personal responsibility for Japanese imperialism.

Though published memories of the Asia Express rarely touched on the train's international relations context and were silent on the causes and consequences of Japanese management of the railway or political control of Manchukuo, they often referenced these questions obliquely through the repetition of imperialist rhetoric about the puppet state. Kubota Masaji, a former section head in SMR's Engineering Department, explained the reasons behind the decision to develop a high-speed passenger train between Dalian and Xinjing in terms that were standard in 1930s Japan, resting on spurious claims that Japan was helping the people of Manchuria: "Manchukuo, born as a new state on March 1, 1932, developed in one great stride after another, and with the need for close connections between Japan and Manchukuo in the areas of politics, military, and economics, the increase in the volume of traffic intensified daily."[61] This explanation emphasizes the successful development of Manchuria under Japanese control and touches on close bilateral connections without reference to violence or the asymmetrical power relations between the two countries. The general idea that the SMR had laudable goals and intentions pervades these recollections. Thus, the silences in these memories are, in a sense, clear statements in support of a particular view of Japan's imperial history that absolves SMR employees and the Japanese people as a whole of guilt.

Former Asia Express driver Mizukoshi Hiroshi's recollections reflect a more extreme version of imperialist rhetoric: "We SMR people, who wished for the welfare of the myriad Asian peoples and dreamed of the great task of Asian development, day and night poured our hearts and blood into the development of continental Manchuria, and in the undeveloped plains, cultivated a transportation network with over 12,000 kilometers of rail and 100,000 kilometers of roads, and under the burning sun, under the moon with cold wind piercing our skin, our spirits never bending or bowing, fulfilled the responsibilities of the great SMR mission for continental rail transportation."[62] Mizukoshi, who seems to have enthusiastically embraced the wartime rhetoric of "raising" Asia, persisted in seeing the train as beneficial to China even in his present day. He wrote that the "heroic" work of the SMR laid the basis for the development of the People's Republic of China, and that, continuing into the future, "the great spirit of the establishment of SMR will be received by future generations of Communist Chinese youth."[63] Though few reminiscers went as far as Mizukoshi, looking back from a historical distance, others praised the legacy of the Asia Express and its place in world railroad history. For instance, Kubota called it "a wonderful start to the high-speed era in world transportation history."[64] Significantly, an awareness of the Western gaze lingered, coloring memories of this train's importance to SMR propaganda of empire. Several pieces mention the American newspaper delegation that was so impressed by its test ride in 1934, and Kubota also recalled the inclusion of the streamlined engine in Loewy's 1937 book featuring locomotives from around the world.[65]

Reflecting the speakers' interests and concerns, recollections by former SMR employees focus almost exclusively on the physical infrastructure and machinery of the train: locomotives, cars, tracks, and other equipment. Passengers rarely make an appearance, but one exception is an anecdote told by Tayama Kazuo, former head of the Planning Section in SMR's Engineering Department. Asked about the air conditioning system, Tayama recounted an embarrassing mishap when SMR president Matsuoka Yōsuke happened to be on board with an army commander, and the air conditioning broke down.[66] As this story indicates, when people did appear in public reminiscences, they tended to be Japanese and were often connected with SMR. Similarly, the few published photographs that include a human element feature either SMR employees posing with their creation or well-heeled passengers on the platform, about to board, seated in the first-class observation car, or enjoying a formal meal in the white-tablecloth dining car. The impact of such a tight focus is to erase the Manchurian people from the picture. Hence the reader sees only the image of a Japanese contribution to Manchuria's development. There is no fighting, no suffering, no usurpation of land, property, or sovereignty in memories of the Asia Express.

Occasional hints of the railroad's role in Japan's imperialist past sneak into these largely happy memories, but they invariably come from someone other than the reminiscer himself. Sometimes this takes the form of a nod to (or complaint about) the historiography of empire. For instance, Mizukoshi, the driver who evoked the SMR employees' passion for the welfare of Asians, lamented that twenty years after Japan's "bitter experience called 'defeat,'" the Japanese people had forgotten about the SMR's achievements, leaving only the disgraceful label of "aggressive war" remaining.[67] Of course, he did his best to remove that label with his insistence that the legacy of the SMR was still helping the people of China. In contrast, the editor of the special issue in which Mizukoshi's article appeared linked the trains of the SMR to Japanese aggression through his explanation of the timing of the special issue. Its late-summer publication was inspired by the numerous historical events that took place in or around July: the July 1931 Wanpaoshan Incident, which the editor notes helped bring about the Manchurian Incident; the June 1931 killing of Japanese army captain Nakamura Shōtarō while on an espionage mission in northern China; and, of course, the July 1937 Marco Polo Bridge Incident, the spark that led to full-scale war between Japan and China.[68]

The introduction to the 1966 episode of the *My Shōwa History* (*Watashi no Shōwa-shi*) television program devoted to the Asia Express made some effort to put Japanese viewers (and later readers) into the shoes of Manchurians. The narration specifically mentions one point that comes up frequently in both contemporary and postwar materials: the name of the train written on the rear of the first-class observation car in hiragana, the Japanese syllabary. Asking what that must have looked like to local Chinese speakers, the narrator draws a comparison to Occupation-era Japan, when American forces put names like "New York" on trains dedicated for use by the American military. "So if we remember the deep emotion with which we looked at that, we can well imagine" what the local Chinese people must have felt in viewing the Japanese writing on trains running through the heart of Manchuria. The show's writers certainly intended a broader critique, suggesting that it was "a strange thing for 'Manchukuo,' whose motto

Critique of historic trains

was 'Harmony Among Five Races,' to run its trains with Asia written in hiragana. Perhaps it expresses the true substance of Paradise of the Kingly Way and Harmony Among the Five Races."[69] Calling attention to the emptiness and hypocrisy of the slogans of Japanese imperialism seems to be pointed toward refuting rosy memories like Mizukoshi's about earnest efforts to help the people of Manchuria.

However, after making that point, the narration takes a sharp turn to the standard encomia, stating that the Asia Express surprised the railroading world in its day and is now remembered with nostalgia and pride. And when the interviewer attempts to draw out his guests on the issue of the train's written name, he does so with a vaguely worded and easily redirected question. He asks former SMR engineer Kubota, "What about 'Asia' written in hiragana on the back of the observation car?" Kubota simply launches into the usual stories about how the name was selected (through a popular contest) and the design of the insignia alongside it, then veers away from the question entirely to comment on the luxurious appointments of the first-class observation car's interior, comparing its elegance to the new bullet train.[70] The show's efforts to inject a critical view into this discussion of the past thus have little impact, and the overall effect is a celebratory display of nostalgia for the days of empire.

Fifteen years later, the effort to mount a critique of SMR's railway imperialism was still ongoing. In his 1981 study of the organization, railway historian Harada Katsumasa suggests that historians were just beginning to connect the history of railroad development in Manchuria to an understanding of Japanese aggression on the continent. After explaining the technical characteristics and propaganda uses of the SMR's "poster train," as well as the implementation in postwar Japanese trains of technologies pioneered on the Asia Express, he notes that historians "are in the process of showing the real form of the 'fifteen-year war' that began with the Liutiaohu Incident as a war of aggression in China. Mantetsu is increasingly being brought into that violent maelstrom."[71] In a 1994 book, technology writer Maema Takanori still focuses primarily on the technical success of the Asia Express, but he places it in the context of a long history of railroad imperialism in northeast China and even assigns some war guilt to the super-express itself. Given the train's prominence in books and pamphlets promoting emigration and travel to Manchuria, Maema asserts, it was instrumental in expanding the population of Japanese farmers in Manchuria, who would soon be abandoned by the Japanese army to suffer undefended through the Soviet invasion.[72] This still limited the definition of "victims" to Japanese nationals, but it took a significant step toward connecting the history of the SMR and the Asia Express to that of imperialist aggression and expansion.

BULLET TRAIN ON THE HOME FRONT

Like the Asia Express, the "bullet train" planned in the 1930s was tied closely to the Asia-Pacific War and Japan's wartime imperialism. But the postwar memory of a "phantom railroad that was never realized"[73] differed dramatically from that of the celebrated Manchurian train in both tone and character. In the absence of photographs of "gallant" locomotives or tales of exciting rides to distant cities, the primary object of public memory was the limited construction completed before the project was suspended: the digging of several new tunnels. This was a story of frustrated ambitions rather than striking success. But retellings rescued the project from the ignominy of defeat by showing how the work that went into the unfinished line ultimately contributed to the achievement of the New Tōkaidō Line. This narrative of continuity contained within it the suggestion that wartime suffering and loss had not been in vain.

Rather than blocking out painful and unpleasant memories, as reminiscences of the Asia Express generally did, historical and fictional accounts of the wartime bullet train homed in on the hardships involved. It was, however, a very narrow view of hardship, limited to those suffering within the shores of Japan, with no consideration of the non-Japanese victims of the imperialist aggression that formed the context and reason for the new line.[74] Memories of the Asia Express and the phantom bullet train both became tales of railway prowess and technological expertise, but in the latter, instead of a conquering hero, the protagonist is more like a scrappy underdog, retreating temporarily to lick its wounds and eventually re-emerge triumphant in a new Japan. Turning a critical eye to the way this story was told, both at the time it was being planned and in retrospect, elucidates one way in which the bullet train acquired meanings beyond its planned technical function of moving people quickly between Japan's major cities. In contrast to the story of the Asia Express, in this case, the aesthetic function of infrastructure reverberated backward in time, renewing interest in a story of failure that became, in the retelling, just the first step in a great success. Instead of a country in the midst of losing a war being forced to abandon a project that had been deemed essential to victory, the story is refocused on determined railway officials struggling consistently against both the military and the devastation brought by war and ultimately succeeding in building the transportation backbone of the new Japan. This story depicted the New Tōkaidō Line as the culmination of decades of progress and technological development and, in doing so, helped redefine postwar Japan from a defeated nation to one that had struggled through adversity and persevered for ultimate victory.

As in the case of the Asia Express, the seeds of postwar narratives about the phantom bullet train were planted in the 1930s. The wartime image of the line was one of national pride and optimism. In the context of total war and construction of Greater East Asia, high-speed rail promised to make a key contribution to both of those endeavors. The project's planning was closely tied to military needs and to the rhetoric and realities of Japanese imperialism in northeast China and Korea. Though it was intended to carry both civilian and military passengers and freight, it was publicly promoted and gained government approval largely as a tool of military power, to expedite the movement of war material and troops to the front in China. And although it would traverse only the "inner territory" of the Japanese empire, as a means of speeding up transportation to a key point of connection with Korea, Manchuria, and China, the line was imagined as an important infrastructure for tightening the relations among them, thereby helping to build the unified East Asian region that was central to Japan's imperialist ideology.

The demands for transport created by the war and policies toward colonized Korea and Manchuria were a motivating force behind the plan. The idea to build a standard-gauge railway across Japan's largest island of Honshū was not new. Though Japan's early railroads all used a narrow gauge, railway officials soon began to sense that they had made the wrong choice on that point, and proponents of a change to the international standard gauge of 1,435 millimeters (often referred to in Japan as "wide gauge" in comparison to the existing narrow-gauge railways) gained ground beginning in the late nineteenth century. The debate over whether to adopt the wider standard gauge went on until around 1919, with the fortunes of either side rising and falling with changes in government and the shifting stance of the military. Two decades after the gauge debate seemed to have ended, the crisis of war and the pressures of empire generated the political will to approve the costly project.[75]

The escalation of the war in China in 1937 increased demand for transportation, so raising capacity in order to avoid bottlenecks became a primary consideration for planners in the Railway Ministry. The role of war and empire in the bullet train project is clear in a report by the Trunk Line Research Department. Created in July 1939 to investigate ways to respond to increasing pressures on existing rail capacity, the department delivered its final report to the railway minister in November, then publicized its findings in the June 1940 issue of the government periodical *Weekly Report* (*Shūhō*). The report begins by explaining that military transportation, rising production, and intensifying communication with the continent were building pressure on the railway system. The existing Tōkaidō and San'yō trunk lines, linking the country's six major cities and harbors, together held the greatest political, industrial, cultural, and military importance. In order to preclude an imminent bottleneck on these lines, the report concludes, the government must build a double-tracked standard-gauge high-speed line enabling trains to cover the nearly 1,000 kilometers from Tokyo to Shimonoseki in nine hours at a maximum speed of 150 kilometers per hour.[76]

The report adds a note of urgency due to the importance of the projected line to mobilization of resources and labor. This reflects a central tension in the plan—a long-term undertaking projected to take fifteen years—between the realities of the construction process and the immediacy of the need to expand transport capacity in order to fight the war. The authors emphasized not only military needs, but also the effects of imperial expansion, arguing that in order to establish the New Order in East Asia, the nations of the region must be considered as a single whole, and they envisioned the completed new trunk line as "the great artery of Asian development connecting Japan, Manchukuo, and China."[77] This idea of the bullet train as a central infrastructure of empire entered the public imagination through the media, which used similar descriptors in articles about the planned line, including "Asian development bullet train," "trunk line for the new East Asia," and, hewing more closely to the original report, "the great artery of Asian development."[78] Promotion of the bullet train as a tool for building Greater East Asia drew it into visions of Japan's imperialist project.

Such notions were bolstered by a contemporaneous plan to build an undersea tunnel connecting Japan to the Korean peninsula. Completion of the world's first undersea tunnel, the Kanmon Tunnel between Shimonoseki and the northern tip of Kyūshū, though it was of a vastly smaller scale, suggested the possibility of an undersea rail connection between the Japanese islands and the continent.[79] Public confidence that the massive project could be completed was linked discursively to imperialist goals by the promise that it would bring the countries of northeast Asia closer together. Newspapers reported that after completion of the bullet train line, trains would be able to pass through the two undersea tunnels to the continent, making it possible to travel from Tokyo to Manchuria and China in a single day without getting off the train, thus realizing the "Asian development dream" of fast, convenient travel from Tokyo to Beijing and points in between.[80] In the public imagination, these two planned works of transportation infrastructure would physically unify the separate countries of Asia. The importance of this connection was also significant in the decision for standard gauge, as the shared gauge would enable a continuous journey from the home islands of Japan, through Korea, and into Manchuria and China. It was the continuity of the journey, in addition to its speed, that differentiated this project from existing connections via ferry. As transportation official Tachibana Jirō noted in 1940, it would be important to match all standards with railways on the continent, so that passengers could "board a sleeping car in Tokyo, cross the sea while asleep, and wake up the next morning in Beijing."[81]

In addition, the empire had a concrete impact on the bullet train plan. Specifically, the SMR's Asia Express served as model, standard, and test subject. The initial plan produced by the Trunk Line Research Department in 1939 (and approved by the Railway Ministry in January 1940) hints at the function of the colonies in shaping transportation infrastructure within Japan proper in its specification that the level of construction should be equal or superior to trunk lines in Korea and Manchuria.[82] The Asia Express had been designed and built by Japanese engineers, but since it was located in

Manchuria, it represented the purportedly independent state and the SMR, not Japan's Railway Ministry. Therefore, the fact remained that since 1934 the periphery had bigger, stronger, faster, and more luxurious trains than any operating in the metropole. "Correcting" that backward imbalance appears to have been one consideration informing the bullet train plan. The Asia Express also impacted the technical development of the bullet train, as Japanese railway engineers visited Manchuria to study the famous train and run tests using its tracks and locomotives. In September 1940, railway engineers from Japan, Manchuria, and China gathered in the puppet-state capital of Xinjing to celebrate "the birth of the bullet train" with an international technology competition. This event had the dual purpose of sharing expertise on such specialties as rolling stock and tunnel construction, while also coordinating plans to connect the line with transportation facilities on the continent.[83] In subsequent months, "top engineers" from Japan's Railway Ministry made several visits to SMR's Technology Research Institute and the Dalian Railworks to conduct tests using the Asia Express and its track, reported in Japan as "the first tests for realization of the bullet train."[84] This kind of connection in the popular imagination between the Asia Express and Japan's high-speed line in development was spurred on by the images of the photogenic Asia that frequently accompanied articles about bullet train plans. These helped Japanese readers to picture the bullet train and created a visceral connection between the built system in Manchuria and the one planned for Japan. In fact, the bullet train was meant to resemble the SMR's star train not only in its standard-gauge rails but also in many of the specifications of its rolling stock. Plans were for full streamlining—the locomotive, the end carriage, and connections between individual cars would all be shaped to minimize air resistance—and it would match one of the major boasting points of the Asia Express with air-conditioned cars and airtight windows.[85]

But war had a paradoxical impact on the railway system. Though military needs provided an important impetus for expanding capacity, they also hindered construction by hogging both manpower and resources. By January 1943, the bullet train route had been decided, land purchases were under way, and construction had begun on a few key tunnels, but the project was almost immediately hampered by shortages of men, materials, and money, as dwindling resources were directed toward the front. Rhetorically and conceptually, the project was part of the military effort. But while railway engineers urged each other on with exhortations to "win the transportation war," planners struggled to overcome shortages and begin construction of an infrastructure deemed "essential to this decisive war."[86] Though promoted as crucial to military capabilities, railroad construction was less pressing than the fight itself.

Yet even as the war's voracious demand for material and human resources brought a stop to construction, the bullet train maintained its image as critical to the war effort and a building block of Greater East Asia. For instance, in a January 1944 contribution to a regular newspaper section covering the science and technologies of war, Tachibana Jirō, now in the reorganized Ministry of Transport and Communications, exuded optimism. The last active work site on the line had been shuttered, but Tachibana promised that the bullet train would soon carry the hundred million masses of Japan's "inner territory" to the "Greater East Asian transportation center" of Kyūshū and move raw materials from all over Asia to the industrial areas of Japan, which he called the core of Greater East Asian chemical and heavy industry production. He depicted the halt to construction as just a pause, explaining, "The current crisis has necessitated the temporary closure of construction from the New Tanna Tunnel on down, but in due time, the bullet train track will draw a line more than two thousand kilometers long, and this great new artery of transportation will continuously communicate its passionate heartbeat throughout the nation, forming the spearhead of new Greater East Asia construction."[87] Continuation of the imperialist aggression and war that had necessitated the bullet train line precluded its completion, but persistent confidence that the plan would eventually be realized helped lay the groundwork for later connections between the wartime plan and the postwar bullet train.

One prominent link in that connection was tunnel construction. Work on the line began in 1941 with several long tunnels that were expected to be the most difficult task. As a result, when construction was halted two years later, the tunnels were the only tangible remnants of several years of planning and preparation. That particular remainder offered powerful symbolism. In Japan's mountainous landscape, tunnels are essential to building a railway line that is straight enough for high-speed operation: tunneling through mountains eliminated the need for tight curves that would force trains to slow down and allowed for a shorter distance between stations, thus speeding up the overall journey. But given the ground quality where the bullet train line required tunneling—some of which was exceedingly hard, some dangerously close to underground aquifers, some insufficiently stable—they were extremely difficult to dig. The New Tanna Tunnel between Atami and Kannami (see map 1, p. xii), for instance, was to go through an active volcanic area, where vents of pressurized gas and water could cause accidents, and workers had to cope with high volumes of water seepage. This would be built alongside the existing Tanna Tunnel, which had taken sixteen years to build (from 1918 to 1934), during which accidents caused the deaths of sixty-seven workers. So the Railway Ministry's ability to build long tunnels through often challenging terrain was a prerequisite to their plan's success and a demonstration of their technological leadership in that area. Perhaps more important to creating lasting memories, difficult construction was the source of dramatic and inspiring stories of determined individuals confronting the dangers and obstacles that nature laid in their path and pushing themselves to the limits of human strength for the sake of the nation. The combination of these factors made tunnels particularly useful for making claims about Japan's superior technological capabilities and national spirit both during and after the war.

The planned Tokyo-Shimonoseki line would include several times the total tunnel length as on the existing Tōkaidō and San'yō Lines.[88] Railway ministry tunnel specialist Arima Hiroshi conveyed the importance of tunnels for high-speed rail to a mass national audience in his contribution to a 1942 collection of articles commemorating the seventieth anniversary of Japan's first railroad. Arima had recently supervised two major projects, the difficult and deadly Tanna Tunnel and the undersea tunnel that was about to open beneath the Kanmon Straits separating Shimonoseki and Kitakyūshū. Placing recent technological development into a frame that was typical of such railway anniversary celebrations, Arima explained that Japanese railroad technology started out half a century behind other advanced industrialized countries, but through "the spirit of those who sacrificed and the passion of the engineers who preceded us," tunnel builders were able to overcome the many obstacles in their path to reach the present state of international leadership in tunnel technology, which the bullet train would demonstrate.[89] Tunnels were also celebrated in popular culture, for instance in the 1942 play *Tanna Tunnel* (*Tanna Tonneru*), "a gritty salute to the brave men who died . . . digging Japan's longest railway tunnel."[90] This discourse gave tunnels symbolic meaning in public perceptions of the Japanese railway system, itself a symbol of Japan's power and modernity.

While the Kanmon Tunnel was extolled as an exemplar of Japan's tunnel technology leadership as the first to run underneath a body of water, the New Tanna Tunnel would be notable for its size: the country's largest, about one hundred meters longer than the existing Tanna Tunnel and built for the wider gauge. Because of the difficult geological challenges, combined with its size and the almost unbearable heat and humidity inside the tunnel as it was being dug out, it quickly became notorious as the most challenging part of bullet train construction. Reporting on construction progress framed the hard work and suffering of the workers as a demonstration of their "spirit" and an important part of the war effort. For instance, the *Yomiuri* newspaper predicted a speedy completion of "the greatest tunnel of the century for the bullet train, the great artery of Asian development," in spite of the difficult construction involved, because the Railway Ministry had "mobilized the best of our engineers, who are worthy of boasting to the world."[91]

Even after construction was halted in 1943, the idea of the tunnel continued to play a role in promoting the war spirit. The completed section of tunnel became a "classroom born of war," in a changing education system that transformed students into soldiers and schools into military training grounds. A 1944 *Yomiuri* article described the change, contrasting the peacetime position of schools, characterized as standing aloof from the rest of society, with the new wartime approach to education, in which there was "no distance between the fighting front and schools," and the mission to destroy the enemy "burns fiercely in the students' chests."[92] As a concrete example of this trend, the article described the Atami Tunnel Technical Training School, which opened in 1942 using the construction site of the New Tanna Tunnel as the work-study classroom. The program aimed to instill a "spirit of construction," described in a pamphlet distributed to all students: "The ideal of construction engineers is not fame and fortune, nor easy living. Just as military men seek fortune in war, [the engineer wants] to be blessed with the great construction project of a lifetime, to participate in difficult construction work."[93] That spirit was amplified by the student-workers' daily confrontation with the potential danger of their labor. Every morning and evening, as they entered and left the tunnel entrance, they would pay respects at a memorial to their "martyred predecessors" (*junnan senpai*) above the entrance to the original Tanna Tunnel.[94]

Accompanied by a photograph of shirtless student-workers wielding hand tools, the article vividly describes the barely tolerable humidity within the tunnel, compounded by the students' spirit and passion heating them from within. It explains that after two years of work and study, graduates of the program could immediately be deployed in construction of railway tunnels or military projects, and concluded, "The well-known brilliance of our tunnel science group, about which we can boast to the world, will soon be increased by the diligent studies of these half-naked students."[95] The unfinished tunnel thus became a site for boys too young to enlist to contribute to the war effort. By presenting the tunnel workers as state heroes and martyrs, articles like this one fixed the New Tanna Tunnel in the public mind as a symbol of the efforts and sacrifices for war that were taking place on the home front. That image would return two decades later.

NOSTALGIA FOR WARTIME HARDSHIPS

Because the project was abandoned soon after construction had begun, the bullet train left behind no photographs of powerful locomotives or travelogues to rekindle fond memories of riding along at high speeds. Memories of the failed project that were sparked by construction of the New Tōkaidō Line, therefore, had a very different tone than recollections about the Asia Express. But they shared the tinge of nostalgia, a feeling that is not limited to rosy memories of happy times. Hardship, too, can be an object of longing for something lost, thus shaping memories of the past and understandings of the present. The direct connections between the early development of the line and the eventual realization of a similar plan in the postwar bullet train fostered narratives that placed it (often alongside the Asia Express) within national and global histories of high-speed rail development as the origin of the world's fastest train. In this way, memory of war helped to bridge the chasm of defeat and occupation, unifying wartime and postwar Japan in spite of the dramatic changes that had taken place in the intervening decades. In the context of celebration of the 1964 opening of the new line, emphasis on direct links between the wartime plans, which were deeply imbedded in Japan's imperialist aggression, and the postwar train had the effect of asserting a positive outcome of war and empire.

As Igarashi explains, some postwar recollections of the past helped individuals and society grapple with the trauma of loss by rebuilding connections to the present across the schism of defeat. "Loss was transformed through narrative representations into a sacrifice needed for Japan's future betterment," and postwar success reframed the pain of war as "a necessary condition for Japan's postwar prosperity."[96] Kerry Smith details a prominent and relatively recent example of this in his study of the Shōwa Hall, a Tokyo museum devoted to the period of the Shōwa emperor's reign (1925–1989). The museum's exhibits connect the past to the present by drawing "a picture of hard-working, committed, and stoic civilians struggling against and ultimately overcoming terrible obstacles to create modern-day Japan."[97] He demonstrates that even painful memories of the past are thus reinterpreted as part of a transwar story of struggle toward ultimate success. In the Shōwa Hall, the suffering depicted is solely that of the Japanese people within the confines of the Japanese islands. There is no suggestion of violence committed by Japanese against others or of the reasons for the war that was bringing such hardship.[98] Mariko Asano Tamanoi considers the significance of recollections of adversity from an empire-wide perspective. Interviewing Japanese and Chinese people about their memories of life in Japanese-controlled Manchuria, she finds that some use the past to redefine their present relationship to the Japanese state by remembering their suffering as a contribution to contemporary peace and prosperity.[99] In similar ways, the focus on sacrifices made by those involved in 1940s bullet train construction—specifically, for the New Tanna Tunnel—contributed to a narrative of gratitude to the wartime generation for their contribution to Japan's postwar success, symbolized by the completion of the long-planned standard-gauge, high-speed railway. In nostalgic views, as during the war, infrastructure was held up as a marker of progress.

Relatively little public memory remained of the wartime bullet train, as an unbuilt infrastructure, but books written about the New Tōkaidō Line during and soon after its construction generally include some reference to its wartime precursor. For instance, JNR's own report on construction of the line highlights both conceptual and tangible connections between the wartime and postwar projects, positioning the original plan as an idea that finally came to fruition twenty years later. In a foreword to the report, JNR president Ishida Reisuke opens with a nod to the line's wartime history, calling it the realization of the people's long-standing wish for a standard-gauge trunk line that "could be seen in the bygone bullet train plans."[100] The first chapter begins with an introduction of those plans as the foundation of the New Tōkaidō Line. By starting the narrative in the 1930s, JNR emphasized the benefits that planners gained from the research and preparation that had taken place for the Tokyo-Shimonoseki line. The debates over gauge, station placement, and other basic elements, as well as practical measures like land surveys and purchases, shaped the basic contours of the plan and allowed for a quick start to construction by abbreviating the process of selecting and preparing the route.[101]

A more popular history, the aforementioned *One Hundred Years of Speed*, also enumerates the ways that research, planning, and preparation for the wartime bullet train facilitated construction of the New Tōkaidō Line. The book presents the 1930s as the first chapter of that success story, calling the wartime bullet train the "womb" of the New Tōkaidō Line. It bolsters the story of continuity by highlighting the significant role in both projects of Shima Hideo, who had led design of the locomotive and cars for the wartime project and was often called the "father" (*umi no oya*) of the bullet train as JNR's chief engineer. It concludes a chapter on the "vanished bullet train" (*kieta dangan ressha*) by suggesting that it had not vanished at all but rather formed a key moment in a decades-long and finally successful effort to realize high-speed rail travel in Japan by using the wider standard gauge: "The concept of the epochal bullet train bit by bit collapsed and ended in a dream, but the devotion of wide-gauge advocates was realized magnificently twenty years later in the New Tōkaidō Line."[102] This shines a flickering light on the wartime bullet train. It is a short-term failure but a long-term success, a mirage concretized into real infrastructure that, by the time the book was published in 1969, was carrying over a hundred million passengers every year.

However, while it credits the earlier plan for the boost it gave to the long-term project of standard-gauge high-speed rail, *One Hundred Years of Speed* does not celebrate the war and empire that motivated its approval and implementation. It acknowledges the planned line's context and connection to the ideologies of the time, for instance by noting that 1940, "the year in which celebrations of the 2600th year of the Imperial reign took place, is the first year in which the bullet train plans took a step toward realization."[103] Noting the nationwide commemoration of Japan's foundation by its mythical first emperor draws attention to the emperor system within which those plans were made. The book also explicitly blames the war for delaying the project, concluding that because the original plan was supposed to be completed by 1954, had it not been impacted by the war, Japan would have had a bullet train a decade earlier and extending from Tokyo all the way to Shimonoseki, rather than terminating in Osaka: "Yet again," the author laments, "we have to regret the vacuum of the wartime era."[104] As Smith has shown, narratives of war that focus on the suffering of the masses of Japanese people reinforce a "victim consciousness," in which blame is placed squarely on the military or even the war itself as the agent of suffering, and the Japanese people are its victims.[105] By reminding readers that the bullet train, which was a source of national pride for the "new Japan" of the 1960s, had a long transwar history, narratives of continuity reconnected postwar Japanese society to its imperialist past, but they also helped people avoid a sense of responsibility for that past by putting the blame for violence and oppression on faceless entities like "the war" or "the military." In bemoaning the setback to high-speed rail caused by the war, One Hundred Years of Speed similarly transformed the phantom bullet train from a tool of war to its victim.

A different kind of reminiscence looked back not to the war but rather to the railway martyrdom that had loomed over the students of the Atami Tunnel Technical Training School: the construction of the first Tanna Tunnel. In August 1958, as plans for the bullet train were being debated, Yomiuri columnist Takagi Takeo connected the loss of lives in that construction to the prospect of work on the New Tanna Tunnel being resumed. His column was part of a series intended to view the present in terms of a painful past. The first installment, published on the thirteenth anniversary of surrender, began "Today is a day that we ninety million [Japanese] people cannot forget, but on the other hand do not want to remember, and that many people probably think they want to forget."[106] Concerned about growing support for nuclear weapons among Japanese leaders, in spite of the history of the atomic bombings of Hiroshima and Nagasaki, Takagi urged his readers not to forget the lessons of the past by relabeling "defeat in war" (*haisen*) as "the end of the war" (*shūsen*).[107] The concept behind this series was that people must remember the aspects of history that they most want to forget, so that the nation does not repeat its mistakes.

It was in this context that Takagi considered the new bullet train plans, viewing the immediate future in the light shed by the events of the past. In spite of the foreboding headline "Skeletons and the Shinkansen," the column initially seems to echo popular excitement with a note of enthusiasm, beginning with the simple statement, "The 'dream super-express' is great."[108] But as Takagi goes on to list the details of the planned railway system, his praise quickly begins to sound rote and even dismissive. "Tōkaidō Shinkansen, wide-gauge. Only three hours between Tokyo and Osaka. Really great. Vibration prevention. Noise control devices. Top speed of 250 kilometers per hour with air conditioning installed. It's all great."[109] One begins to sense a "but" on the horizon, and it comes with discussion of the New Tanna Tunnel. The column, it turns out, was prompted by a gruesome discovery. Preparatory geological investigations using the original Tanna Tunnel brought to light the remains of one of the individuals who had perished in its construction. Of the men whose lives were lost, remains of only half had been recovered, leaving more than thirty corpses presumably "plastered into the concrete walls of the tunnel."[110] With the set phrase "thousands die to raise one hero to fame" (*isshō kō natte bankotsu karu*), Takagi raises the specter of additional sacrifices for the new high-speed line. He implores his readers not to forget the dead who were already sacrificed for speed and to consider those yet to be sacrificed for the "dream super-express."[111] In this case, connecting the contemporary bullet train project to an earlier incarnation produced a story that, rather than rehabilitating the past through connection with a glorified present, called into question the state's plans for the immediate future by tying them closely to a disastrous past.

Even such a negative example, however, shows that tunnel construction could make for a powerful story. Tunnels form a key part of bullet train memory because of their importance to the project. They are particularly useful in retelling the past and present for two reasons: first, the danger and difficulty of the task are ingredients for good drama; and second, the challenge of blasting through mountains could be used, as it was during the war, to demonstrate not only the "spirit," but also the technological capabilities of those behind it. As the most challenging part of construction and one of the sites at which work completed for the wartime train was resuscitated for use on the new line, the New Tanna Tunnel was ideal as the crux of stories linking past and present.

One example that reached a mass audience is a two-part episode of the television show *Bullet Train* (*Dangan ressha*), which tells an emotional story about the hardships and losses suffered by those who worked on the first stage of tunnel construction. This show reveals a different aspect of memory than either the celebratory remembrances of the Asia Express or the straightforward acknowledgment of earlier contributions to a current project. By framing a wartime story within opening and closing scenes of the 1962 celebration of the tunnel finally piercing through the mountain, the show suggests that the struggles depicted were all worthwhile. This story constituted the second and third episodes of the ten-part series that aired on the national broadcasting station (*Nippon Hōsō Kyōkai*, or NHK) from August to October 1964, in the weeks leading up to the new line's grand opening. Taking advantage of popular excitement and interest in the bullet train as its opening day approached, the series told semifictional stories about the process of planning, building, and preparing to operate the line, each episode focusing on a particular challenge or hardship that had to be overcome in order for the system to succeed. The show was compelling and exciting for reasons beyond just the timely and dramatic story line. Reportedly the first serialized television drama to use color film, it was visually captivating.[112] In addition, the episodes were filmed in part on location at bullet train construction sites, lending the stories an air of realism. One newspaper review cited the New Tanna Tunnel episodes as exemplifying the challenges and benefits of having actors become part of the construction site, which gave the drama the realistic feel of a documentary.[113] The show thus brought its audience viscerally into the moment of the story, adding to its emotional impact.

The story tugged on viewers' heartstrings to instill a sense of gratitude for the hard work and suffering of the wartime generation, which is portrayed as having contributed directly to the growing prosperity of the 1960s, symbolized in this narrative by the bullet train. It achieved this effect through three steps. First, it established the initial planning for the Tokyo-Shimonoseki line and the eventual completion of the New Tōkaidō

Line as a single transwar project. Second, it positioned the national railway alongside the Japanese people as the passive victim of the army and of the war itself. And third, it connected the characters in the story to the entire population of Japanese people old enough to remember the war years through depictions of widely shared experiences.

The very inclusion in a television series about the New Tōkaidō Line of a story about the wartime bullet train implies that the two railway projects, separated by a span of fifteen years, should be seen as a single continuous endeavor. The prominence of this story within the series—as a double episode, it constitutes one fifth of the entire run—communicates the importance of the wartime precursor to the nearly finished project. The story focuses on the New Tanna Tunnel as the most well-known and tangible link between the two projects. And while most of the action takes place between 1941 and 1945, it is bookended by scenes of an important postwar milestone. The first episode begins with two main characters leading a group of workers to blast through the last section of rock separating the two ends of the tunnel and tearfully wishing that their wives were there to see it, before flashing back to the spring of 1941. We return to the same event for the closing scene of the second episode, thus enveloping the wartime story in a postwar frame.[114]

Within the action of the story, too, alongside references to the construction being part of the war effort, characters make statements that foreshadow its postwar revival. For instance, one of the main characters, Construction Site Assistant Director Hayashi, gives an impassioned speech about the importance of the high-speed trunk line they are building. When one of the workers suggests that the war threatens the line's completion, Hayashi responds: "I don't know about war, but I know about trains. The era of running wide-gauge high-speed rail will absolutely come to Japan. The existing Tōkaidō and San'yō trunk lines are already saturated. In order to conquer this situation, the only thing is high-speed rail, and that has to be wide-gauge."[115] And the New Tanna Tunnel, he adds, is essential to that future line. Regardless of the war, Hayashi tells his co-workers (and the audience at home), the New Tanna Tunnel and the high-speed rail line for which it is being built will eventually come to be. This attitude persists even after he learns that construction has been suspended and the remaining workers have been drafted into an "occupied territories construction corps." Hayashi and his wife, Taneko, discuss the fate of the 2,000-meter section of the tunnel that they have already dug, which will fill with solfataric clay unless it is constantly maintained. Taneko suggests that it does not matter if the tunnel fills up, since the bullet train has been suspended. Hayashi responds, "Yes, for now . . . but when war ends, the bullet train will definitely run! I can see it. At that time, I want this pilot tunnel to be useful for it."[116] Taneko, begging him to return alive, promises to maintain the tunnel while he is gone. These scenes work to extricate the rail project from its association with war and imperialism, as Hayashi's insistence that the reasons for wide-gauge rail transcend momentary war needs provides a past for the bullet train that is stripped of its military impetus.

The rest of the episode shows Taneko and other workers' wives and mothers, joined later by tunnel workers who have returned from the front after sustaining injuries, desperately keeping the tunnel from filling up with clay while fighting hunger and exhaustion. Their success and the accuracy of Hayashi's predictions are revealed in the final scene depicting the moment in September 1962 when the two digging crews that had been working toward each other finally met to create a single tunnel. The scene opens with an advertising balloon reading "Celebration: New Tanna Tunnel Drilled Through." Hayashi peers into the tunnel that has been blasted open and explains in a voice-over that when the construction group returned to Japan in the spring of 1946, it seemed the maintenance of the tunnel had been for naught; the context of dire poverty was not promising for a bullet train. "But our suffering bore fruit thirteen years later! Construction of the Tōkaidō Shinkansen! The New Tanna Tunnel construction started from the 2,000-meter pilot tunnel that we dug and that our wives maintained!"[117] By connecting the wartime and postwar bullet train projects, the show's writers promoted the idea that the work completed in the context of total war mobilization contributed directly to a shining symbol of Japan's postwar economic, industrial, and technological success, which viewers would get to enjoy when the new line opened in just six weeks. But the final credit to the wives who led the effort to maintain the tunnel against the elements serves to write over the story of warfare among nations with a tale of the human struggle against nature. This depoliticized the history of railway development in order to color the technocratic state as the continuous force behind rational progress. It also reinforced the gender roles of the emerging enterprise society by presenting women as productively partici-pating through support of their husbands.[118]

Even as it linked the bullet train to the wartime period, the show also helped cleanse it of any sense of war guilt or responsibility by represent-ing the railroad bureaucracy, employees, and the line itself as victims of the military and the war. It presents the standard explanation that mili-tary needs were the catalyst for the project, but the action increasingly sets the railroad workers in opposition to the army. Chatting with a fellow train passenger, Taneko boasts about her husband's important work on the tun-nel, explaining that it is key to the bullet train, which is needed because the Tōkaidō Line is not sufficient to transport men and materials for the prolonged war, echoing the frequent wartime claim about the line's raison d'être.[119] But throughout both parts of the story, there are constant remind-ers of the war's negative impact. In one scene after another, there is a drum-beat repetition of comments on the insufficiency of labor supplies, and characters share news of one worker after another being drafted.[120] More explicit opposition is shown in an interaction between two railroad men, Hayashi and Construction Director Yasuda, and some nameless army of-ficers checking on the progress of tunnel construction. Told of delays due to the difficult ground quality, the officers disparage the Railroad Minis-try's "underdeveloped" technology. As soon as the officers leave, Hayashi and Yasuda mock them and complain, "It's awful to be blasting this tun-nel for guys like that."[121] This comment makes it clear that the military is the motivating force behind the line, but it also draws a sharp line between the military support that pushed politicians and bureaucrats to approve the construction and the real aims and motivations of the railroad men.

Throughout the story, occasional scenes depicting moments of daily life that were part of the wartime experience of a large portion of the Japanese population would have evoked a more general nostalgia for the past and, importantly, provided a point of direct connection to the story for many viewers. After the last of the railroad workers are called up for military ser-vice, the first episode ends with a scene of soldiers receiving a celebratory send-off from the townspeople.[122] Around the same time, the female char-acters gather together to go food shopping, departing in the early morn-ing before the men are even awake, and head to an area several miles from the construction site because they think they may be able to buy potatoes daikon radishes in that area. They are later shown dividing up among themselves their small reserve of rice, declaring that they are like a single family.[123] Finally, in an experience shared by all too many, the worker-turned-soldiers' wives wait anxiously for their return, and one learns that her husband has died.[124] By connecting the particular events of these characters' lives to shared experiences of the wartime and early postwar years, the show generalizes its message, suggesting that, like the railroad families of the story, the sacrifices of all Japanese people—the lack of food, the backbreaking work taken up by those remaining at home, the loss of loved ones—helped to build the strong industry, economy, and society of contemporary Japan.

In stories like this, public memory of the phantom bullet train seems simultaneously to praise and condemn the forces behind wartime railway development. Such works thus speak in the same "double register of doubt and complicity with the state" that art historian Namiko Kunimoto identifies in the avant-garde artist Nakamura Hiroshi's 1964 painting of a steam locomotive barreling along with such speed that it is flying off the rails. In Kunimoto's reading, the train equally represents national power and disaster by forming a link between past and present that "connects the disastrous consequences of imperial Japan's ambitions to the rhetoric of new state power embodied by the postwar emphasis on infrastructure and economic growth."[125] Reminiscences of the phantom bullet train seem to share this kind of double register. They praise and celebrate the fast, sleek train of the New Tōkaidō Line, but at the same time, by connecting it explicitly and tightly to the disastrous war, they potentially (and sometimes purposefully) inject an element of doubt.

Progress often evokes nostalgia. Even amid the anticipation and excitement surrounding the bullet train's opening day, some felt an unmistakable longing for what was being lost. In a dialogue with the president of the Railroad Friends Society Kyoto Branch, published in a local newspaper just days before the bullet train's first run, Kyoto Station Master Takai Hitoshi commented that although the bullet train was the culmination of JNR's efforts and captured the attention of the nation, people of his generation felt nostalgia for the smoke-billowing locomotive. He lamented that already "the shift to electric from the steam locomotive drowned out one flavor of travel."[126] The bullet train pushed the nation even further from that romanticized past. A few days later, a reporter for the *Asahi* newspaper on the first Osaka-bound Hikari echoed the key talking points about the dream super-express, which was "at long last no longer a dream": like so many others, he noted the world record speed, safety, and smooth ride, and predicted a new era for railroads. But sliding into Osaka, he searched in vain for the feeling inspired by the steam whistle and worried that modernization would make rail travel into something businesslike and unemotional.[127] These individuals focused on what was being lost, but public memory of past trains inspired by the new railway system also forged new connections between past and present, collectively shaping understandings of both.

Just as the bullet train became a symbol in contestation of Japan's imperial past, it was also taken up by those who sought to influence its global future. The preceding chapters have all focused on Japanese views of the new line's impact on their own individual, local, or national identities. But with the development of technological assistance and industrial displays as diplomatic tools, the bullet train was taken up as a means to change international perceptions of Japan. The meanings attached to the bullet train in the context of Japanese society were projected abroad through aid programs, trade negotiations, and international cultural events. Reception outside Japan was, if anything, more complex, making the bullet train as difficult to wield as a tool of diplomacy as one of domestic development policies. The next chapter considers the successes and failures of bullet train diplomacy.

5   Technology of Cultural Diplomacy

IN THE FALL OF 1963, American organizers of the New York World's Fair, set to open the following spring, sent a flurry of letters and telegrams to Tokyo in an effort to convince Japanese officials to ship a bullet train car to New York for an elaborate display to be included in the fair's Japan Pavilion.[1] Months later, protectionists in Congress worked just as hard to keep Japanese trains out by adding a "Buy American" clause to legislation funding mass transit projects.[2] Around the same time, state and federal government agencies were sending representatives to Japan to study the line under construction with an eye toward developing similar projects along the corridor from Boston to Washington, D.C.[3] This conflicting set of actions reflects the varied responses to Japanese efforts to mobilize the bullet train as a tool of technology diplomacy.

Japanese government agencies energetically promoted their success in building the world's fastest train to an international audience, part of a wider effort to use displays of Japanese technologies and industrial products to change negative views imprinted during the war and early postwar years. JNR invited people from all over the world to view the bullet train under construction, participate in test rides, and receive training in railway engineering. They also distributed promotional materials—including actual components of the system—in order to introduce it to a more extensive

global audience than the relatively few who could visit Japan. Such efforts
did not always work as intended. While Japan's bullet train diplomacy enjoyed some success in changing international perceptions of Japan, it was
constrained by the preconceptions and expectations of Japan's diplomatic
partners.

By the late 1950s, views of Japan were changing but still colored by
the recent past. Postcolonial governments in Southeast Asia increasingly
looked to Japan for economic and technical assistance, and the reparations
negotiations required under the peace treaty were finally completed by 1963.
But animosity and distrust cultivated by wartime domination hindered relations with other Asian countries, which were further complicated by the
ideological tensions of the Cold War. In the United States, wartime views
of the Japanese people as irrational, duplicitous militarists were quickly replaced during the Occupation period by less awful but still derogatory stereotypes of the subservient geisha or obedient student of American democracy.[4] Shock that the Japanese could produce the formidable Zero fighter
turned to disdain for the cheap trinkets and shoddy knockoff products that
entered postwar U.S. markets labeled "Made in Japan." The punitive attitude of the early occupation soon gave way to support for Japanese recovery, and the Eisenhower administration focused on building a mutually
beneficial security and economic partnership, including technical aid, yet
maintained an imperious attitude toward its ally.[5] While certainly an improvement over wartime perceptions, neither official nor popular views of
Japan suggested a position of equality in the bilateral relationship. But by
the early 1960s, new exports began to inspire admiration for Japanese industrial capabilities, and the Kennedy administration began working toward an "equal partnership."[6] In this context, Japanese government and industry leaders sought to feed these trends, hoping to improve relations and
boost exports by incorporating the display, transmission, and promotion of
technology into diplomacy.

Measures of technological achievement have long played a motivating
role in international relations.[7] In this sense, improving perceptions of Japanese technology were about more than just increasing exports. Government
and business leaders aimed to build trust in and reliance on Japan among
developing countries in order to build regional relationships that would
bolster Japan's economic growth and regional position. And they strove

to erase persistent American images of Japan as quaint and backward, not only in order to increase demand for their industrial products but also to move toward a more balanced diplomatic partnership. On a global scale, demonstrations of Japanese industrial and economic success constituted a kind of ammunition in Cold War battles over the best model for development. The bullet train (and rail systems more generally) figured in efforts by various government agencies to improve Japan's international position and strengthen bilateral relations with both developing and advanced industrial countries by putting Japanese technological capabilities on international display. However, such efforts were constrained by two overlapping factors: preconceptions about Japan and the particular function of technology in international relations. In Asia, development assistance was welcomed but could potentially create an impression of economic invasion. In the United States, war memories and desires for an exotic Japan worked from opposite directions to hinder efforts to update timeworn stereotypes by shaping displays and coloring their reception. Because technology was closely linked to both military potential and economic power, Japan's former enemies and current industrial competitors were liable to perceive threats in displays of technological prowess. The use of the bullet train as a diplomatic tool demonstrates the promise and pitfalls of Japan's showcasing technological accomplishments and capabilities as a diplomatic strategy.

Domestic and international political circumstances of the 1960s combined to bring technology to the forefront as a tool of diplomacy. As prime minister from 1960 to 1964, Ikeda Hayato responded to Cold War tensions and tumultuous domestic politics in the wake of the 1960 mass demonstrations over renewal of the U.S.-Japan Security Treaty by adopting a "low posture" foreign policy. His successor, Satō Eisaku, continued this general approach of avoiding friction while asserting Japan's economic position in the world.[8] As a result, Japan's role in the international community centered on economic activities, including rising exports of electronic consumer goods, advanced machinery, and industrial components. This approach to foreign relations was reinforced by the legacies of World War II. Japan's postwar constitution, designed to ensure that the nation would never again threaten world peace, prohibited the use of force to resolve foreign policy problems, while a wariness of Japanese power among other Asian countries restrained forceful leadership on controversial issues. This combination of factors tightly restricted the use of either military force or economic pressure in international relations.

While constraining some modes of foreign policy, the legacies of war also opened up space for other methods. For instance, the Japanese government's efforts to rebuild relationships with Southeast Asian countries damaged by wartime aggression focused on reparations agreements. Negotiated in the context of the postcolonial nation building of the 1950s and 1960s, these agreements included development projects using Japanese equipment and expertise. The structure of reparations agreements evolved into the Japanese government's system of Official Development Assistance, a key part of its regional foreign policy in the postwar period.[9] This fit into broader Cold War initiatives, as the United States pressed Japan to take on a greater role in technical assistance to developing countries in Asia. Helping Asian neighbors improve their railroads is an example of the ways in which, as anthropologist Ashley Carse argues, "Infrastructure was Cold War politics by other means," becoming the "material manifestations of Cold War geopolitical struggle."[10] Japanese reparations and development assistance exemplify the ways in which foreign policy was closely tied to domestic industrial interests, so much so that even a diplomatic matter as delicate as war reparations negotiations was viewed in terms of Japan's industrial and economic needs. Diplomats hoped that technical aid would create feelings of respect, admiration, and trust toward Japan by associating it positively with the infrastructures of development.[11] Technology thus entered into Japanese foreign policy in part because it was a resource that Japan could provide and that was needed by its diplomatic partners, but also because it was seen as potentially cultivating the much more intangible resource of favorable attitudes toward Japan (what today might be referred to as "soft power") in countries where such sentiments remained scarce more than a decade after the war's end.[12] This was part of a two-pronged use of technology as a tool of diplomacy: toward developing countries in Asia, the Japanese government promoted its role as a model of development and source of technical assistance, while in relations with industrialized countries like the United States, technology diplomacy focused on creating desire for Japanese products, reducing barriers to trade, and promoting a more equal diplomatic partnership by building respect for the Japanese people and nation.[13]

he 1960s emphasis on technology in Japanese foreign policy making
inued a long history of equating technological achievements with na-
al identity, security, and welfare.[14] The Cold War intensified technology's impact on international relations, in part because it was so central to
the legitimacy of each side in the conflict.[15] Novel elements for Japan at that
particular historical moment included both its international position and
its technological capacity. Achievements like building the world's fastest
train cast Japan in a position of global leadership, and private organizations
joined with government agencies to use displays of the bullet train to bolster cooperative relationships, raise the nation's global status, and change
negative stereotypes.

 But efforts to use technology to strengthen Japan's international friend-
ships were complicated: displays of technology inspired respect, but they
also aroused fear, anger, and resentment. The economic resurgence and
military potential associated with Japan's technological prowess were both
sensitive points in its foreign relations at the time, as memories of military
conflict or wartime domination still shaped Japan's relations with its former World War II enemies, and Japanese products began to compete with
those of Western industrialized nations in electronics and heavy industry. In the case of Japan's closest diplomatic and trade partner, the United
States, the high-technology image that Japanese leaders hoped to convey
clashed with the particular friendly images of Japan propagated after the
war, which focused on traditional material cultures like woodblock prints
and ceramics. Falling outside the popular view of Japan, the image of a
technological power risked interpretation through a darker lens. The successful promotion of Japanese technology to both Western powers and developing countries in Asia, therefore, potentially brought along some negative by-products.

The bullet train's diplomatic function reflects a multifaceted shift in Japan's international position in the 1960s, precipitated by economic growth
and technological development. In 1963 and 1964, Japan gained recognition
as a developed economy in the General Agreement on Tariffs and Trade
(GATT) and the International Monetary Fund (IMF), giving up allowances
made to developing economies. Japan also joined the Organisation for Economic Co-operation and Development (OECD) in 1964. In these and other
venues, members of international organizations for economic cooperation

began to recognize the importance of full Japanese participation.[16] These changes collectively signaled Japan's shift from a country requiring economic and technical assistance to one with the means to provide it. It was the same surge in economic growth that motivated the decision to build the bullet train, and railway technology played a role in both changing global perceptions of Japan and the shift in its international role, with two new rail systems—the New Tōkaidō Line and the monorail connecting Haneda Airport to the city center—completed just in time to take advantage of the global focus on Japan afforded by the 1964 Tokyo Olympic Games.

THE WORLD'S FASTEST TRAIN AND
THE SOFT POWER OF HARDWARE

Japanese officials took advantage of the attention that came from building the world's fastest train by mobilizing the new rail system for public diplomacy initiatives, including technical assistance to developing countries and impressive displays aimed at other industrial powers. JNR and other government agencies invited foreign representatives to Japan to see the new line under construction, sent engineers and advisors abroad, and exhibited the bullet train through displays of machinery and other materials. Development of the railway industry dovetailed with foreign policy concerns through its incorporation into foreign aid programs. At the same time, Japan still continued to receive some technical assistance from the United States. The temporal overlap of these contrasting roles of aid recipient and provider is exemplified by the fact that the bullet train, which would become a global model for high-speed rail, was built with an $80 million World Bank development loan.[17] The extensive public praise for this infrastructural development rarely mentions the external funding that made it possible. It is significant that this fact was mostly swept under the rug in public relations, as it may have detracted from the image that government leaders sought to promote through the project, that of an economically and technologically advanced nation.

Even as it was receiving substantial international support for the bullet train, JNR participated in repositioning Japan as a provider of technical aid by training railway experts from other countries and giving technical advice on construction and operation. For example, in 1962 (when the line was still under construction but was already well known in the railroading world), responding to requests for technical advice on various aspects of railroad development, JNR sent experts to work with national railways in India, Ceylon, Pakistan, Thailand, Burma, Malaya, the Republic of China, Saudi Arabia, Syria, Argentina, and Panama. In the same year, foreign railway officials visited JNR or sent engineers to study railway engineering for several months at a time from both developing and industrialized countries, including South Korea, Germany, the Philippines, the United States, Burma, Sudan, Egypt, and France.[18] The surprising inclusion of France, a railway leader, and the United States, whose technical experts in various fields had recently been in the opposite role of providing assistance to Japanese specialists, highlights the function of railway expertise in helping to create a new position of international leadership for Japan. JNR itself used this program to burnish its reputation in the global railroading community, boasting in a new English-language periodical published by the Japan Railway Engineers' Association (JREA), "It gives the JNR engineers great pleasure that railway engineers all over the world take a deep interest in JNR."[19] Such programs built on efforts pursued through the 1955 Bandung Conference, bilateral reparations agreements, and other avenues of economic and technical cooperation, to create and expand new frameworks for public diplomacy, fostering positive feelings for Japan by developing opportunities for mutually beneficial interactions between a Japanese government agency and individuals of various nations. These ostensibly apolitical activities helped Japanese leaders navigate the difficult path between two of the most important goals of postwar foreign policy: pursuing a leadership position in a left-leaning postcolonial Asia and maintaining full-fledged membership in the anti-Communist camp in the Cold War.[20]

A 1965 article in the JREA journal described JNR's overseas technical cooperation more fully, detailing five categories of assistance. First, the dispatch of experts around the world (mostly to Asia), often in coordination with the Ministries of Foreign Affairs and International Trade and Industry and the government's Overseas Technical Cooperation Agency, had been generally rising over the previous decade. Second, the training of personnel from overseas railways (mostly from Asia, but also Europe, the Middle East, and Latin America) was also on an upward trend. Third, JNR regularly accepted requests from overseas railways or governments to inspect rolling stock for export from Japan "in the belief that by so doing it will contribute to the promotion of friendly relations with foreign railways."[21] Fourth, they also engaged in technical cooperation through exchange of information and data in response to "request after request from overseas railways for advice on technical problems," as well as setting up overseas offices in New York and Paris.[22] Finally, agency representatives partici- pated in many international conferences. Noting the reversal of Japan's po- sition in global railroading since its nineteenth-century beginnings, the ar- ticle concludes that, having benefited in the past from foreign guidance, JNR aims to contribute to other countries' economic and industrial de- velopment: "JNR is now able to do something in return for what overseas countries have done for it in the past, particularly from the point of railway technique."[23]

The journal itself was a product of such efforts at international outreach. Given the difficulty of bringing people to Japan to see the bullet train in person, JNR publications—from glossy pamphlets to specialized explana- tions like those in the JREA journal—were an important way to show off the new rail system, along with events, films, books, and magazines de- signed to "introduce" Japan to the world. The journal was born around the same time as the bullet train, its first volume published in July 1959, just as concrete planning for the line was getting under way. The content would interest only a specialized audience of railway engineers, but the creation of a technical publication in English (with French and sometimes Spanish ar- ticle abstracts beginning in 1960) reflects the sense that Japan was in a po- sition to make a valuable contribution to global expertise in such areas. The bullet train was a frequent topic in the journal's early years, with informa- tion about every aspect of its development and operation.

The technical training of foreign engineers and other cooperative ac- tivities of JNR show how three exigencies of foreign policy overlapped with JNR's development of railway technologies: the need to develop ex- port markets in Southeast Asia, efforts to repair diplomatic relations in Asia, and the Foreign Ministry's search for new modes of diplomacy in the postwar context. The first of these three elements is related to the state of the railway vehicle (rolling stock) industry in Japan. The demands of early postwar rebuilding efforts brought a rise in production of new trains, and by the mid-1950s, the industry suffered from excess capacity and was eager to develop overseas markets. Japan was not yet exporting the bullet train technology itself, but it contributed to a positive impression of Japan's technological capabilities in the industry that could help promote exports. The latter two elements were interconnected, as the legacies of World War II and complexities of the Cold War context led the Japanese government to an emphasis on development projects as the central pillar of its relations with former victims of Japanese imperialism.

Of course, the bullet train was conceived as the solution to the practical problem of an overcrowded railway line, not as a tool of diplomacy or international one-upmanship. But officials involved in foreign relations latched onto the idea of using the train's aesthetic functions—the glamour of its speed and design, the sheen of its high-tech electronic communications and automated controls—to promote a new image of Japan as a technological powerhouse. International displays of the bullet train mostly involved bringing people to Japan to see it, study it, and ride it. This happened in part through regular diplomatic structures, such as the United Nations Economic Commission for Asia and the Far East (ECAFE). Having initially been excluded from the UN as an "enemy state," then prevented by Cold War politics from joining for several years after the occupation ended, Japan was admitted in 1956, after reaching an agreement with the Soviet Union to normalize diplomatic relations. The government was eager to use UN activities to rebuild Japan's international prestige, and the UN's auxiliary organizations, several of which Japan was able to join before being admitted to the UN, were a good venue for this, removed as they were from the Cold War tensions that often constrained action in the General Assembly and Security Council. Japan joined ECAFE as an associate member in 1952, acceding to full membership two years later. This organization was particularly useful in Japanese efforts to rebuild connections with the countries of Asia, as its fundamental goal was to promote economic reconstruction and development in Asia, in part by strengthening economic ties within the region.[24] Thus, ECAFE's activities resonated with the basic thrust of Japanese policy in the region.

Under ECAFE auspices, railway officials, experts, and diplomats from fifteen countries visited Japan in April 1963 to study the New Tōkaidō Line, then still under construction, giving JNR a chance to impress the world with its technological prowess through these representatives. Organizers initially planned a meeting of rail engineers, but broader interest in the bullet train led them to expand it from a technical meeting to a large-scale observation tour that included government and industry leaders from the two superpowers (reported to be the most enthusiastic observers), major European railroading countries, and developing countries of Asia.[25] The group included not only ECAFE member countries (Australia, the Republic of China, France, India, Indonesia, Malaya, Pakistan, the Philippines, the Soviet Union, Thailand, the United Kingdom, and the United States), but even two nonmember nations (Canada and the United Arab Republic), and several UN officials.[26] The U.S. State Department arranged for participation by eleven delegates, including individuals from private industry, the New Jersey state government, and relevant federal agencies.[27] Participants enjoyed a ride on the completed test track, experiencing both high speeds and precise, computer-controlled automated braking. They examined support structures for elevated tracks, specially designed rails with insulated joints, a rail bridge and tunnel, and various aspects of the computerized automated train control system. And they visited construction sites, new station facilities, and JNR's Railway Technical Research Institute, which had developed aspects of the system that were essential to safe, comfortable high-speed operations. Several delegates extended their stays beyond the weeklong organized tour in order to continue studying JNR's organization and technology.[28]

Judging by Japanese news reports and assessments from American participants, the study week was quite successful as a promotional effort. The major Japanese daily newspapers reported breathless admiration from participants throughout the week. The *Asahi* reported that participants exclaimed "Unbelievable!" in reaction to the automated braking and uttered cries that rose almost to a scream when the speed on the test ride surpassed 250 kilometers per hour. The scream was one of excitement, not fear, as they also described a pleasant ride: "There was very little shaking or noise. It's a magnificent technology."[29] The *Yomiuri* noted that the group as a whole credited JNR with helping the global railroad industry to combat the argument that it was a "sunset industry," losing out to road and air travel.[30]

Members of the American delegation were impressed by the new line as a great achievement not only of engineering but also of political and technical leadership. The State Department approached the event with hopes that some of the innovations of Japan's new line could help solve transit problems at home; a press release announcing participation in the planned study week specified that American representatives would pay particular attention to economic, service, and especially safety features that could be incorporated into domestic rail systems and regulations, noting that the highly industrialized and densely populated Northeast Corridor in the United States and the Tōkaidō region in Japan shared similar transportation challenges.[31] A few months later, the American delegation's report gave the new line high praise, calling it a "'super railroad' in every sense of the term."[32] Specifically, the report noted that Japanese research had brought about the necessary technological developments in various areas, such as track, structures, equipment, signaling, communication, and operating methods. Asserting that the new line "will be a source of pride and prestige for Japan," the report reversed the U.S.-Japan technology relationship that had held for nearly two decades, during which Japan had benefited from American technological assistance. In the case of high-speed rail, it stated, other countries, including the United States, could benefit from Japan's achievements.[33]

The report thus explicitly placed Japan in the position of world leader, stating that JNR's "most important contribution is in providing the leadership to proceed with such an outstanding project in the face of so many unknowns. They had the courage to do what the rest of the world just dreams about doing."[34] The American representatives' positive assessment reached a wider audience through a *New York Times* editorial, which cited a report on the study week by the New Jersey State Highway Department as a potential source of inspiration for mass transit projects in the New York area. The editorial heaped praise on the train's "dream run" at high speeds in cars of "ultramodern design, ventilated and air-conditioned" with engineering features that "represent the latest in railroad technology."[35] While Americans were beginning to admire Japanese light industry's skill in miniaturizing consumer electronics, the bullet train gave them a glimpse of Japan's strength in heavy industry.

Soon after the line was officially opened, the American press echoed the notion that the bullet train was inverting the aid relationship between the United States and Japan. A *Washington Post* editorial (reprinted as a "Guest Editorial" in the *Chicago Tribune*) gushed about the new train as "an example of what a railroad could and should be."[36] The details of speed, style, and safety features provided a stark contrast to the state of American rail, which led to the writer's tongue-in-cheek suggestion: "Of course, Japan is an advanced country, but perhaps the government in Tokyo could be induced to send a peace corps and a team of technical assistants to the United States to help us modernize our eastern seaboard lines."[37] This was obviously meant as a joke, but it is the ring of truth that made it funny. In fact, some very influential Americans thought Japan did have something to teach them about rail. After seeing the nearly complete New Tōkaidō Line, Rhode Island senator Claiborne Pell proposed a similar rail system be built along the Boston-Washington "Megalopolitan Corridor," then pestered President Lyndon Johnson until finally winning his support for the idea.[38] In April 1965, two engineers from MIT, commissioned by the federal government, completed a survey of the New Tōkaidō Line to see what lessons it might offer. They, too, had high praise for Japan's accomplishment and envisioned a similar system as the future of mass transportation in the United States.[39] The High-Speed Ground Transportation Act, which passed with bipartisan support in September 1965, prompted private railway companies to develop designs for power-saving, streamlined trains and resulted in a research and development program for high-speed passenger transportation along the Northeast Corridor.[40]

As this suggests, just as rail-related and other technical assistance projects helped to rebuild diplomatic relations with less developed countries, displays of the bullet train and other Japanese technologies helped to change broader international perceptions of Japan, contributing to efforts to promote its industrial exports. One program targeted a particular audience: the Technical Tourism Program was aimed primarily at expanding Japanese exports by attracting businessmen who might be convinced by firsthand observation of Japanese industries and personal contacts to begin purchasing the products of Japanese industry. Inspired by a French program of industrial tourism, leaders in Japan's tourism industry began promoting similar activities in the late 1950s. These efforts were initially aimed at tourists from Southeast Asian countries, but with an increase in overseas travel by Americans in the early 1960s and an expected influx of foreign visitors from around the world for the 1964 Olympics, planners expanded their target audience to advanced industrialized countries.[41]

The program opened early in 1965 with the cooperation of several institutions and a wide variety of industries. It was organized by the JTB with the support of the Ministry of Foreign Affairs, the Ministry of Transportation, the Japan External Trade Organization (JETRO), the Ministry of International Trade and Industry, and the Japan Chamber of Commerce and Industry, as well as other tourism organizations, and involved forty-one leading business and industrial concerns and numerous smaller companies.[42] The idea behind the program was to create flexible tours combining famous scenic, historic, and cultural sites with visits to leading industrial facilities. The bullet train was a centerpiece, not only as an attraction but also, of course, because it enabled visits to businesses in both the Tokyo and Osaka areas within a brief period. A *New York Times* article on the program noted that "besides the 'Bullet Train,' . . . other innovations in fast travel have added to the comfort of touring Japan."[43] A widely distributed informational pamphlet about the program explained that most tourists saw only one aspect of Japan, so this tour would give them a more complete and accurate view by showing them the modern industry that shapes daily life, including "one of the best subway and train systems in the world."[44] Rich color photographs showed the imperial palace, Mount Fuji, Osaka Castle, and famous temples and shrines, along with a bird's-eye view of Tokyo's bustling Ginza shopping district, ships in the Yokohama Port, and the bullet train pulling into Tokyo Station. This tour, the brochure promised, would show both the expected traditional culture of Japan and its modern technological wonders.

Though there is no record of how many business travelers took advantage of this opportunity or how it affected their views of Japan, the program's extensive cooperation among government agencies and private interests demonstrates the broad-based effort to leverage Japan's cultural attractions and natural beauty in order to exhibit the country's industrial strength and modern technology toward a combination of goals. In the words of a JTB representative, "The tours will offer not only sound, tangible proof of Japan's amazing economic development, but they will help spur business, promote even better relations with nations around the world, and provide visitors with a keen insight into the new Japan, as well as a charming touch of the old, ancient ways."[45] This summation of the program provides a glimpse of the ever-present pairing of old and new that pervaded Japan's public diplomacy ventures in the twentieth century but also reflects a shift in emphasis that characterizes the 1960s in particular: Japan's modern technology is clearly the focus, on which visitors will receive "keen insight."

Of traditional culture, on the other hand, there will be only a "touch." The bullet train was ideal for this approach, demonstrating contemporary Japanese design and technology while providing both a platform from which to view Japan's natural beauty and speedy transportation to sites featuring traditional architecture or other industrial showpieces. The bullet train later became a selling point in advertisements for regular commercial travel. For instance, a 1968 series of advertisements for a package tour included riding the bullet train past Mount Fuji in its description of the highlights of a nine-day tour of Japan, which focused otherwise on cultural and historic sites.[46]

That people involved in more generalized efforts to promote Japan's international image perceived the bullet train's potential impact on perceptions of the country is evident in the numerous English-language publications that touted its wonders. One example of this outward-looking presentation is an English-language introduction to Japanese culture and society called *Meet Japan: A Modern Nation with a Memory*, a title that captures the conscious pairing of old and new that pervaded Japanese image-making throughout the twentieth century. Commissioned by Dentsū, Japan's largest advertising company, as an official souvenir of the Tokyo Olympics and subsequently republished for a broader audience, the book was part of the general effort to introduce a "new Japan" that was characteristic of 1960s public diplomacy.[47] Amid chapters on Japan's ancient capitals, castles, and arts, as well as new expressways and major industries, a section on the bullet train begins by noting that railway experts throughout the world had praised the New Tōkaidō Line as the "Railway of Tomorrow," reinforcing the perception beginning to take hold in the industrialized world that Japan was already producing futuristic technology only imagined in other countries. But it was not for the sake of speed alone that the train was built, it claimed; on the contrary, JNR "cared naught for making a useless challenge to speed, nor for building vain-glorious vehicles, but instead wanted to construct the most highly modernized railroad possible to meet Japan's demands for mass transportation."[48] That disclaimer, however, was followed by boasts about speed, safety, and amenities, stating "these trains run at a speed never before attained in world railway history."[49] The description of the train conveyed the desired image of Japan as pushing the cutting edge of technology, operating a train of the future in the present day. But at the same time, the inclusion of traditional arts and castles alongside technology and industry emphasized the nation's long history and cultural heritage. This injected an element of techno-Orientalism into old Orientalist tropes of Japan as a storehouse of an exotic past.[50]

Newspapers and magazines helped bring the bullet train to a broader American audience. In an article for the *New York Times*, science writer Lawrence Galton gushed over the smooth ride and computerized control by a central headquarters in Tokyo. Given the relatively backward state of rail transportation in the United States, Americans describing the Japanese train often tried to capture its speed by comparing it to faster modes of transportation; Galton reported passengers' views "that traveling on it is like riding a rocket-powered sled on ice and is the closest thing to jet travel on the ground."[51] An article in *Popular Science* announced that the new federal program created by the High-Speed Ground Transportation Act would bring "exciting new high-speed trains" designed on the model of the "remarkable Japanese system" but also warned that costs might prevent the United States from attaining that goal.[52]

Putting the bullet train into a global context, technology writer Joseph Gies included "the fastest train" as one of a dozen "wonders of the modern world," along with the world's biggest bridge, longest tunnel, tallest building, and other impressive achievements in infrastructure, urban planning, industrial production, and weaponry. Like the American participants on the ECAFE tour, Gies acclaimed JNR's foresight and will in building the line, which he presented as a model for the United States and elsewhere: "It almost could not be done, and many people said it was impossible, but the Japanese National Railways undertook to do it for the Tokyo-Osaka stretch, and what they accomplished may well spell the future for the even bigger Boston-Washington metropolitan corridor, not to mention several other places in the world in the next half century."[53] He also praised its safety features, such as the automatic controls that kept it within designated speed limits, and its comfortable, elegant interior. But he put a particular spin on his evaluation, noting that "a brimful whisky jigger set down on the window sill does not spill one drop."[54] Though even steam trains—such as the Pennsylvania Special that set a record of 127 miles per hour (about 204 kilometers per hour) as early as 1905—had reached similar speeds in special runs, "in that swaying, plunging, clickety-clacking

[handwritten marginalia: Americans couldn't even comprehend the bullet train]

Pennsylvania Special . . . there is no record of anybody setting a brimful jigger of whisky on the window sill."[55] Nor, he continued, was there another example of a tightly packed schedule of high-speed trains running over 300 miles (about 500 kilometers) with split-second precision. "Continuous fast operation on a crowded schedule, with nonspill comfort and absolute precision—that, rather than rate of speed itself, is the achievement of the New Tokaido."[56] The unspilled jigger of whiskey was apparently a compelling vision for American readers—it also appeared in an enthusiastic review of the train by *New Yorker* magazine's "far-flung correspondent."[57]

This extensive praise suggests that the bullet train might be labeled as a success of technological diplomacy, helping to build a new image of Japan as a source of technical expertise and a model to be followed, rather than a nation needing assistance. In this sense, one result of the bullet train's aesthetic function was that it helped to put Japan in a new position of global leadership. But it was a small segment of American society that would either experience the bullet train or take great notice of reporting on it. Changing broad popular views would require more attention-getting venues.

FROM RICKSHAW TO RAILROAD: JAPAN AT THE 1964–1965 NEW YORK WORLD'S FAIR

JNR rushed to complete the bullet train in time for the Tokyo Olympics in part to facilitate transportation of the expected crowds traveling in and out of the host city but also to take part in the international display of Japanese technology that the event afforded. The thousands of foreign Olympic visitors provided a timely opportunity to show the train off to a global audience. At the same time, JNR specifically targeted an American audience by contributing a bullet train display to the Japanese Pavilion at the New York World's Fair, held over two six-month periods from late April to October in 1964 and 1965.[58] The rail agency cooperated with JETRO (which organized the government exhibit with the support of the Ministries of International Trade and Industry, Foreign Affairs, and Finance) to promote the train and its underlying technologies internationally in connection with the nation's foreign policy goals. This is one example of the ways in which high-speed rail became part of a broader effort to use displays of Japanese technologies to promote exports and change popular views of Japan abroad. But the nature of technology as a point of challenge to American economic interests

ndustrial dominance complicated those efforts by evoking old anxi-
about Japanese competition in global trade.

Americans who saw the bullet train were impressed, and the public enjoyed increasingly high-quality imports from Japan, especially consumer
items like radios and television sets, but they were slow to abandon what
the Japanese press and officials derided as the "Fujiyama/Geisha" image
centered on chintzy representations of Japan's natural beauty and traditional cultures. In Japanese materials, the phrase "Fujiyama/Geisha" is consistently written in katakana (the phonetic system used for foreign words),
suggesting a foreign perspective: officials did not object to images of Mount
Fuji or geisha per se, but rather, as JETRO explained in their report on the
fair, aimed to "wipe away the ignorance and bias" implied by these out-
dated images, still common in American society.[59] And at the same time,
old fears of a militarist Japan still lingered under layers of postwar reim-
aging, ready to bubble up to the surface with any prodding. Responding to
this lag in popular American perceptions of Japan, the Japanese government decided to participate in the New York World's Fair with the ambitious goals of increasing American knowledge about Japan and improving
political, trade, and cultural relations.[60]

JETRO's plan was to display "industrial products that boast a world
standard" and various elements of contemporary society in order to demonstrate Japan's dramatic progress over the preceding century.[61] But the
inclusion in the Japanese area of two buildings planned by a private consortium of businesses, the Japanese Exhibitors Association (JEA), in some
ways worked counter to the government's goals. Interested primarily in
profits, the consortium continued the long-standing practice of world's fair
exhibitors, showcasing both traditional cultures and innovative industrial
products alongside spectacle and entertainment. JEA contributions to the
pavilion catered to American preconceptions and exoticist interests and
thus reproduced a self-Orientalizing identity that was prevalent in Japan
at the time.[62] Government planners, too, made some midstream concessions to visitors' anxieties and expectations that ultimately undercut their
efforts to revise Japan's image, instead cementing misperceptions they had
explicitly stated they aimed to correct. New American images of Japan were
thus co-created by overlapping and competing interests in both countries,
informed and shaped by the function of technology in international relations. In this sense, the fair is clearly one of the "middlebrow institutions"

that Christina Klein identifies as serving to "educate Americans about their evolving relationships with Asia" and create "opportunities—real and symbolic—for their audiences to participate in the forging of these relationships."[63] Pavilion organizers and the visitors who absorbed or distorted their intended messages both participated in reshaping Japanese national identity.

Generations of Japanese leaders, like world's fair exhibitors from around the world, had recognized their propaganda potential as venues for nations to showcase technologies and cultures. The 1867 Paris World's Fair, the first to have a government-sponsored exhibit from Japan, showed how these events could shape world opinion, and over the subsequent century, successive governments used them to present an idealized Japan in order to further particular foreign policy goals.[64] As Japan industrialized, that increasingly involved a combination of ancient traditions with modern innovation; the particular circumstances of Japan's international relations at any given time helped determine the balance between old and new. For instance, in the late 1930s, when tensions were high surrounding Japanese aggression in China, Japan's world's fair pavilions aimed to soften the nation's militaristic image by focusing on unthreatening artifacts of traditional culture. At the 1937 Paris Exposition, organizers chose the theme of "Old Japan" for the national pavilion and displayed contemporary materials only as photomurals of everyday life in the communal Hall of Nations. The New York 1939–1940 exhibit saw slightly more emphasis on modern industry, but the balance still tilted heavily toward the traditional. There, the business-backed Hall of Nations exhibit included a photomural titled "Industrial Japan," but only as the backdrop to a display of Japanese crafts; the Japanese Pavilion resembled a Shinto shrine; women in kimono presented elements of Japanese culture; and outside, a "flame of peace" burned inside a stone lantern.[65] Postwar pavilions continued this approach. By the 1960s, Japan was no longer typically seen as a potential aggressor, but the Japanese exhibit at the Seattle World's Fair, held just two years before the event in New York, still emphasized traditional cultures, with a *torii*-style gate forming the entrance, a large rock garden occupying the center of the exhibit, and roughly equal space devoted to industrial and cultural displays.[66]

For the 1964–1965 New York World's Fair, planners took a new approach. Both organizers and the press called for a more accurate representation of contemporary Japan that would sweep away the "Fujiyama/

na" image.[67] JETRO aimed to achieve this image change by exhibit-
he products of Japanese industry, including the bullet train, in order
resent a technologically advanced industrial power, an ideal partner in
trade and development. The government building's most significant nod to
the past, a massive external stone wall designed by sculptor Nagare Masayuki in the style of a Japanese castle, symbolized strength and endurance,
countering associations of contemporary Japan with the delicate or feminine aspects of its traditional cultures.[68] But this was not the only Japan
on display. The divided responsibilities for the Japanese area between two
groups with similar but not identical goals resulted in mixed messaging.
Of the two buildings sponsored by the JEA, one exhibited the participating
companies' products punctuated by elaborate flower arrangements and was
staffed by young female hostesses dressed in kimono. In the other, dubbed
"The House of Japan," visitors could enjoy Japanese cuisine, tea ceremonies,
and dance performances. This combination of displays had the result of
couching cutting-edge technology within familiar Japanese cultural traditions, a framing that undercut the main thrust of the effort to change views
of Japan by reminding Americans of the very images that the government
organizers hoped to replace. That outcome has much to do with American
preconceptions, perceptions, and expectations, reflecting the importance
of reception in cultural diplomacy and the specific challenges of placing
technology at its center. Visitors wanted to see what a *New York Times* review described as an "exotic Japan" where "kimono-clad beauties ride rick-
shaws under the cherry blossoms while fierce-visaged men hurl each other
around with judo holds."[69] The need to attract attention in order to make
any impression at all meant that organizers could not completely neglect
the version of Japan that their audiences expected and desired. As one JEA
official explained to American reporters, "Since we rely heavily on trade,
we want to show our advanced technology. We will have judo exhibits and
tea ceremonies to satisfy the exotic image of Japan, but we will have computers, too."[70] This candid statement reveals the practical considerations
that shaped the Japanese area as a whole: attract audiences with cultural artifacts and activities matching their expectations, then amaze and delight
them with advanced technologies in many fields.

The government building, however, consistently promoted awareness of
recent developments in Japanese industry. JETRO officials decided that in

order to reduce ignorance of contemporary Japan, the government building would "exhibit various Japanese industrial products that boast a world
standard . . . and while showing how Japan has achieved dramatic progress in the hundred years since the opening of ports, also engage in general introduction and publicity about national affairs in all areas."[71] The
designer of the government exhibit, Kamekura Yūsaku, took to heart the
aim of presenting Japan as an industrial and technological powerhouse.
Critiquing displays at recent world's fairs in Paris, Brussels, and Seattle
for simply throwing old and new together in a jumble, he resolved to try
something different: put only contemporary artifacts on display and reference traditional materials through backdrop images, not so much to celebrate that tradition or suggest any direct lineage to contemporary products as to highlight how far Japan had come in a relatively short time.[72] The
result was an exhibit titled "From 1860s to 1960s: Feudalism to the Edge
of Space." It emphasized the nation's rapid development through the constant juxtaposition of examples of the latest products of Japanese industry
with reproductions of woodblock prints depicting comparable items from
a hundred years earlier, just before the modernization and Westernization
of the late nineteenth century. For example, a print of fireworks accompanied a display of rockets; images of a woman lighting a paper lantern and
a blacksmith at work were placed behind massive electrical equipment; a
drawing of men operating an early modern grinding device stood in the
background of an automatic grinder and other machine tools; a print of
traditional weavers was shown alongside an automated "Jet Loom"; an image of a shipwright contrasted with a model of the huge oil tanker *Nisshō
Maru*; a beautiful woman using an abacus accentuated the high technology
of Fujitsu's FACOM-231 computer; and a courier carrying news on horse-
back was paired with a tiny transistor television.[73]

The bullet train, which began operating toward the end of the fair's
first season, was an ideal technology for these goals. The fact that it was the
world's fastest regularly scheduled train meant that it was undoubtedly at
the level of the "world standard." Photos of the train running through scenic countryside and major cities could show off Japan's natural beauty and
architecture. And since the line opened less than a century after Japan's
first railway system was built under the guidance of British engineers and
with imported rolling stock, its success underscored the speed of Japan's

technological progress. Both the JEA in Tokyo and fair organizers in New York were eager to have the bullet train on display. The fair's liaison in Tokyo, an American named Antonio de Grassi, Jr., reported that the JEA was particularly interested in exhibiting the new train and was ready to underwrite a substantial portion of the costs, if the Transportation Ministry would approve the plan. De Grassi had an exciting suggestion: "Since the train is US standard gauge (the first in Japan), it may be possible to land it at San Francisco/Los Angeles and run it across the US to New York."[74] The Japanese government gave a lukewarm response, so the JEA sought help from the fair's president, New York's "master builder" Robert Moses. Moses dispatched letters to the president of JNR and the Ministers of International Trade and Industry, Foreign Affairs, and Transportation, urging inclusion of a locomotive and car from the New Tōkaidō Line in the Japanese area.[75]

It seemed for some time as if the fair might have a substantial exhibit devoted to the train. A pamphlet produced by the JEA in October 1963, just six months before opening day, even included a description of an elaborate display: "The Super-speed Dream Train, known both nationally and internationally as the best of its kind in the world, and as a victory for Japanese industry and its engineers, is scheduled to be exhibited, with provision for projected moving scenes of the Japanese country-side in order to simulate an actual ride."[76] Since the exciting thing about the train was its speed, the projected scene would perhaps have provided a visceral sense of its significance. However, around the same time that this pamphlet was being produced, the JNR leadership finalized its decision not to send an entire car, citing the expense and likely public backlash against such an extravagant effort in the context of public criticism of the rising costs of construction. However, JNR expressed its desire to cooperate in some way, setting the stage for the more modest exhibits of their new rail system that ultimately appeared in the government building.[77]

Visitors caught their first glimpse of the bullet train in "an elaborate miniature railroad system emphasiz[ing] Japan's railroading skills" and boasting a tiny super-express, which the official World's Fair guide book noted was "the world's fastest express train."[78] A little farther along, they would come across an actual piece of the train: its chassis, or undercarriage, sitting on rails and ties shipped from Japan. This was hardly the most glamorous part of a train lauded for its elegant design, but it was important in terms of the technological developments made by JNR's Railway Technical Research Institute to enable safe high-speed operation by preventing swaying at high speeds.[79] The resulting view of Japan as a nation capable of designing and building a rail system of unsurpassed speed, safety, efficiency, and elegance was reinforced and brought into high relief against a background depicting modes of transportation emblematic of the two ends of the preceding century: a photograph of the bullet train speeding across the countryside alongside an enlarged reproduction of a woodblock print depicting men bearing a palanquin.[80]

This pairing, emphasized by a heading that ran across the side-by-side photograph and print, reading "Transportation: 1960s . . . 1860s," told the story of a country that in just one hundred years had made a technological leap from a place in which most travel was either by shank's mare or in a litter borne by human power to the nation that designed, built, and operated the streamlined, high-speed bullet train. The woodblock print itself might have reminded viewers of an important and recognizably Japanese artistic tradition. Japan, the combination hinted, was more than just transistor radios and portable television sets; it was winning the global race for railway speed, even as it maintained traditional cultural artifacts and practices admired around the world. But the point of the display was clearly the contrast, which emphasized the speed of change. Reviewing the pavilion, Robert Trumbull, head of the Tokyo bureau of the *New York Times* for much of the 1950s and 1960s, noted the combination of old and new (given the title of the exhibit, one could hardly miss it), specifically commenting on "the drawing of a rickshaw behind the stark steel bulk of a set of present-day high-speed railway wheels."[81] Though he mislabeled the palanquin as a rickshaw, we can see that the juxtaposition of tradition and modernity made an impression on at least one influential viewer. While the overall theme and arrangement of the items on display invited comparison between the "feudalistic" past and industrial present, the emphasis was squarely on the latter. The organizers used architecture to convey a sense of entering a Japanese space with the external stone wall. But the objects exhibited were all examples of contemporary technological innovation and design, and the comparative traditional items were relegated to prints on the walls behind them. Planners certainly wanted the world to see Japan's premodern past but primarily as a background to accentuate the speed and distance of its technological progress.

The organizers of the Japanese pavilion claimed to have achieved their diplomatic goals. Drawing about a fifth of all fair visitors, it was one of the most highly attended pavilions in the international area. JETRO boasted that its popularity indicated that they were able to show the general population of Americans the high standard of Japanese technology.[82] Americans who commented on the pavilion seem largely to agree. Trumbull's *New York Times* article reported excited responses to Japan's high-speed cameras, "the world's most powerful electron microscope," and a scale model of "the world's largest oil tanker . . . built by the world's largest ship-building nation, and row upon row of working miniature television sets that demonstrate the skill that has made Japan a world industrial leader."[83] Trumbull's list of superlatives echoes the larger effort, of which the World's Fair exhibit was a part, to force a global re-evaluation of Japan's place in the industrial world.

But public reactions to the first session showed JETRO that their exhibit had missed its mark in ways that reflect the impact of preconceptions on reception of cultural diplomacy initiatives and the specific difficulties of displaying technology. The Japan Trade Center Public Affairs Office in New York, which issued a report on the 1964 season, found that visitors were "generally impressed by Japan's scientific and technological advancement" and gained "a better understanding of contemporary Japan," an impact magnified by a detailed television report on the Japanese pavilion and its exhibits broadcast nationally on NBC's *Today Show* that August.[84] However, it also noted a few points where the pavilion fell short of JETRO's goals. An important one was that signboards were quite small and contained overly detailed explanations, so that few visitors, often exhausted from a long day at the fair, bothered to read them. As a result, the point of certain displays was lost, and fairgoers sometimes interpreted both industrial displays and cultural flourishes in very different ways than organizers intended.

For instance, while almost every visitor noticed the impressive stone wall, for which six thousand sculpted stones had been transported from Japan and assembled on site by specially trained Japanese artisans, the received messages were mixed. Those familiar with Japan recognized it as representing the architecture, history, and tradition of Japanese castle building, but the most frequently noted impression was of a fort or prison. Compounding the feeling of entering a military structure was the sight that greeted visitors as soon as they entered the building: several large rockets. Japan Trade Center analysts found that this combination "obviously resulted in the impression of aggressiveness as well as militarism," in spite of an explanatory (but apparently overlooked) signboard titled "Rockets for Peaceful Use."[85] Even a Japanese visitor commented that the building resembled an armory, and the rockets at the entrance gave a warlike impression, adding that some people thought they must be entering the Soviet Pavilion (though the Soviet Union did not participate in the fair).[86] Much larger American rockets on display elsewhere in the fair do not seem to have caused consternation, but pre-existing feelings about Japan shaped visitors' reactions. Thus, two prominent elements of the government's exhibit meant to enhance Japan's image by showing off a classical architectural style and a cutting-edge technology backfired, instead reminding fairgoers of Japanese militarism.

Americans' preconceptions about Japan also led to disappointment with the industrial focus of the pavilion. The Japan Trade Center report noted, "Many visitors, still cherishing the traditional image of Japan with kimono, flower arrangements and tea ceremony, colorful shrines and cherry blossoms, felt the exhibits of the Government Building lacked so-called typical Japanese touch."[87] Similarly, though Trumbull claimed that visitors would see "kimonos aplenty," as well as "stylized dances performed to the twang of the three-stringed samisen, watch judo and a form of stickfencing called kendo, and sample such native delicacies as raw fish and sake (rice wine)," he also remarked that some visitors were disappointed in the paltry effort to display the exotic Japan they had anticipated.[88] JETRO's own survey confirms this impression: nearly half of the visitors felt the government building was not "typically Japanese" and wished there had been more Japanese culture, arts, and crafts.[89]

Organizers, it turns out, had underestimated the depth and tenacity of Americans' misunderstanding of Japan. Their miscalculation led to other stumbles in the effort to erase the "Fujiyama/Geisha" image, as its very pervasiveness shaped popular reception of their displays. For instance, the pavilion's hostesses were all dressed in kimono, a choice that would be unremarkable to a Japanese audience. But this traditional garb gained a particular valence for Americans due to the ubiquity of the geisha figure in popular representations of Japan. In the years leading up to the fair, several Hollywood films set in Japan centered on geisha. By the 1960s, Americans had already seen too many versions of the same story, and the 1961 comedy *Cry for Happy* was panned for retreading the tired theme of foreign visitors misunderstanding the services a geisha provides.[90] The following year, audiences saw Shirley MacLaine transformed into a supposedly convincing geisha just by putting on the right clothes and makeup in the film *My Geisha*.[91] American television shows about Japanese society also prominently featured these traditional entertainers, and just a month before the fair opened, American Express advertised a package tour of Tokyo with the words "Sukiyaki, geisha girls, 1964 Olympics."[92] Therefore, for many Americans, the unusual sight of women dressed in kimono brought to mind the idea of the "geisha" and its varied associations. Since the realms of technology and politics had long been gendered male in both the United States and Japan, Americans' persistent association of Japan with the feminine figure of the geisha worked against efforts to recast Japan as the equal of the United States in the "male" fields of industrial production and diplomacy.[93]

Americans involved in negotiations with Japanese planners shared these perceptions. In an internal memo, a member of the fair's International Affairs and Exhibits staff described the "House of Japan" as having "a traditional restaurant, geisha girls, 99.9% Japanese."[94] Soon after, Tokyo liaison de Grassi sent an update regarding the restaurant's waitresses, concluding with a comment that "there's a chance to get yourself a geisha."[95] To make matters worse, a miscommunication with the fair's publicity office had resulted in extensive advertisement of the restaurant as a "Geisha House." Organizers corrected the mistake in subsequent materials, but it was echoed in American media outlets, and the error was reported by the Japanese press, causing some negative public relations for the fair.[96] The "geisha girl flap," as de Grassi called it, became something of an international incident: the Japanese consulate intervened with a formal request that the word "geisha" not appear in any fair publications or news releases.[97] In spite of this effort, the tenacity of the image was such that the *New York Times* described the women of the Japanese delegation in the fair's opening day parade as "geisha girls twirling parasols."[98] This is just one example of American preconceptions distorting efforts by Japanese organizers to change them, so that the experience of the fair in some ways reinforced old stereotypes rather than breaking them down.

The fair's winter closure gave organizers an opportunity to correct some of their missteps by making changes to the pavilion for the 1965 season. They may have been particularly sensitive to criticisms about the lack of cultural elements because of the growing perception that the Japanese government was excessively focused on increasing exports. In 1963, a French news magazine reported that President Charles de Gaulle had dismissively referred to Japanese prime minister Ikeda Hayato as "that transistor radio salesman" after his visit to Europe. Though there is some question as to whether de Gaulle actually said such a thing, the assertion captures a general perception of Japan's foreign policy.[99] This stinging insult reinforced for Japanese government and opinion leaders a sense that real international respect required displays that would show the depth and complexity of Japanese society beyond the transistor radios and other consumer electronics that were flowing into foreign markets. An opinion column in the prominent daily *Asahi* newspaper, noting the fuss over whether or not de Gaulle had made the controversial comment, concluded that either way, the fact that the insult was reported and hit home was related to Japan's insufficient attention to cultural diplomacy. Criticizing the Foreign Ministry's emphasis on economic policy, the author called on foreign policy officials, who were using only their "economic minds," to cultivate a "cultural mind."[100]

The food and performances in the "House of Japan" were perhaps a nod toward the "cultural mind," as were the stone wall at the entrance and the Japanese garden on the grounds. But for 1965, pavilion organizers tried to expand the cultural aspect of their display. They resisted pressure from the fair's New York leadership to send examples of Japan's "art treasures" but responded positively to the Japan Trade Center's recommendations for two changes.[101] One was to give the exhibit a brighter, more active feel by having the model train running and giving demonstrations of machinery, and by playing more accessible Japanese folk music instead of the avant-garde work written especially for the event. The other was to soften militaristic images by adding the "anticipated Oriental human touch as well as cultural displays," such as color panels depicting sights like gardens and palaces. The report also recommended revising the rocket display, since "many friendly Americans kindly advised us that rockets primarily reminded them of [the] militaristic side of Japan."[102]

In response, the Japanese organizers removed the largest rockets from the exhibit and moved the smaller ones to a less prominent spot in a back corner. The prime location at the entrance was taken over by a small propeller plane and two cars, surrounded by reproductions of nineteenth-century prints and photographs of Tokyo Bay and the Olympic Games, held in Tokyo just as the fair's 1964 season came to a close.[103] The airplane, developed by Mitsubishi in cooperation with the American company Mooney Aircraft, suggested a continuation of Japanese reliance on American expertise, a comfortably familiar hierarchy for the American public. And while cars would come to symbolize the Japanese threat to American industrial and economic dominance by the 1980s, in 1964, they brought a chuckle of disbelief. One visitor reminisced: "The name 'Datsun' seemed strange, and the car looked tinny and cheap. 'Ha ha ha . . . whaddya know, a JAPANESE car!' people said."[104] At that time, the Toyota and Datsun on display represented a novelty, not a challenge to the American auto industry. Images from the Olympics helped to link the "new Japan" on display to this popular symbol of international peace and amity.

JETRO also added some elements of cultural interest and natural beauty to the railroad exhibit, displaying alongside the model trains a collection of photographs of places that one might visit by rail, consisting mostly of temples, castles, and scenic landscapes. They replaced the photograph of the bullet train behind the chassis on display, which featured a well-lit train in front of a dark, nondescript hilly background, with a new image of the train running in front of Mount Fuji, a more recognizable sight. And in order to encourage visitors to take home a memento of Japanese culture, JEA planners expanded a concession area to offer Japanese handicrafts and gifts in addition to the cameras, watches, and consumer electronics that sold briskly in the first season.[105] The Japanese government had come to the fair aiming to disabuse Americans of their antiquated views with an exhibit that celebrated contemporary Japan by contrasting it with the past. But constrained by popular expectations and lingering wartime antipathy, and incorporating a much larger section organized by the profit-oriented JEA, the Japanese Pavilion inadvertently reinforced some of those images. Given its role in both Japan's military potential and its economic and industrial

challenge to the United States, technology was difficult to wield as a tool
promote respect for Japan without provoking negative associations.

The bullet train, however, was deemed a success in promoting Japan's
desired image change. For instance, in a special issue of a transportation
industry trade journal evaluating the new line, Japan Development Bank
official Sanuki Toshio listed "Change in Foreigners' Image of Japan" as one
of the main areas in which the bullet train had been very effective: "Foreigners' image of Japan was 'the country of Fujiyama and geisha,' but has
now changed to 'the country that built the world's fastest Tōkaidō Shinkansen.'"[106] Supporting his claim, he noted that the line's designers had
won international awards, including the Elmer A. Sperry award, for dis-
tinguished contribution to transportation engineering, and the Columbus
Gold Medal, both in 1966. Such engineering-focused prizes likely garnered
little popular attention, but combined with news reports, popular writings,
and public displays such as at the New York World's Fair, the bullet train
helped to change American views of Japan. Sanuki suggests that the bullet train replaced the famous peak and the geisha as the "object" representing Japanese culture. The "Fujiyama/Geisha" image of Japan that fair planners were so eager to erase took hold of the American imagination as part
of an Occupation-era process of replacing Japan's militaristic image with
a peaceful one. In the 1960s, the U.S.-Japan relationship had transformed
such that Japan's fast trains (and transistor radios) were capturing the
American imagination, if not so much as to put them at ease with such potentially threatening technologies as rockets from their enemy-turned-ally.
However, protectionist legislation shows that even railroad technology implied a potential threat to the United States as representative not of a military but an industrial and economic challenge from Japan.

RAILWAY EXPORTS AND AMERICAN PROTECTIONISM

The World's Fair provided a venue for the tightrope act that the Japanese
government performed in its efforts to promote industrial exports without igniting anti-Japanese sentiment. This was made particularly difficult
by the rising tide of protectionism in the United States. As both electronics and products of heavy industry from Japan began to compete effectively with American-made goods by the early 1960s, American companies
pressed their representatives for protectionist measures, which one State

Department expert called "a series of pin pricks" irritating bilateral relations.[107] The technological and industrial progress that JETRO presented at
the World's Fair, therefore, created a conundrum for policy makers. While
Japanese capabilities raised American interest in a greater Japanese role
within the bilateral Cold War partnership, the same strengths caused tensions in the relationship. This dynamic is visible in top-level discussions
taking place at the time. In January 1965 (coincidentally, during the break
between the two sessions of the World's Fair), Prime Minister Satō, just two
months into his tenure, traveled to Washington, D.C., to meet with President Johnson, about to begin his first elected term as president. At this
meeting, railway exports were one area in which Satō pushed against the
rising protectionist trend in American policy, and Johnson showed flexibility, so as to press for Japanese agreement on issues that were more important to his administration, such as Japan's role in shared defense, anti-
Communism, and the prevention of nuclear proliferation.

That said, railway technology itself was not a primary concern for either
leader. During two days of meetings on January 12 and 13, Satō and Johnson discussed major foreign policy goals and tough sticking points in bilateral relations, such as continued American control of Okinawa and the Bo-
nin Islands, Japan's relations with South Korea and China, policies toward
Vietnam, and trade frictions in textiles and civil aviation. But at a moment
when the success of the bullet train was inspiring Americans to dream of
high-speed rail in their own Northeast Corridor, the potential for the export of Japanese railway equipment to the United States was on the list of
topics for discussion. It was a key point in both Satō's foremost goal for the
meeting, which was to push for American trade liberalization, and in Johnson's efforts to appease Japan on trade in order to focus on pressing issues
of national security. State Department planners proposed greater openness
to Japanese imports for mass transit projects as one of just a few points on
which the United States could offer Japan some "good news," in spite of the
broader protectionist atmosphere.[108]

The electronics industry was the site of the most highly visible influx of
Japanese products, as consumer electronics found their way into more and
more American homes, and Japanese semiconductors were proving to be
a cost-cutting component for American electronics manufacturers.[109] But
top-level government negotiations focused on bigger items. Behind closed

doors with the prime minister, Johnson (a staunch free-trader) scoffed at business leaders' anxiety over electronics imports, showing off the three Sony televisions permanently switched on in his office, tuned to the major networks.[110] More directly relevant to these meetings were the inroads being made by Japanese heavy industry, given the context of a resurgence of domestic support for the "Buy American" policy. This policy, encoded in a Depression-era law, institutionalized an advantage for domestic producers in purchases by the federal government. But with American industry's dominance of the global capitalist economy in the period from World War II to the early 1960s, the promotion of nationalist purchasing fell out of favor. The 1933 Buy American Act was temporarily suspended during the war and then significantly weakened by a 1954 executive order. While there were pockets of support for campaigns to promote consumption of domestic products over imports, the politics of free trade mostly prevailed, shaping the policy goals even of major labor organizations, whose leaders reined in anti-import proclivities of their rank and file, based on the assumption that keeping the world's markets open to American goods benefited workers more than closing the U.S. market to foreign goods. But voices of dissent against the free trade mainstream gained volume over the late 1950s and early 1960s, often with a racist tinge familiar from wartime anti-Japanese campaigns. As American labor historian Dana Frank argues, the cooperation of organized labor with free trade ideology in the 1950s set the stage for a resurgence of Buy American policies and explosion of anti-Japanese protectionism in the 1970s and 1980s.[111] By the time Satō was planning his visit to Washington, specific regulations created in response to challenges presented by Japanese suppliers were reviving the moribund Buy American approach to federal purchasing.

In the early 1960s, two controversies surrounding Japanese bids for federal government contracts publicized the success of Japanese heavy industry and rekindled discussion of the costs and benefits of a Buy American policy. In April 1960, as Japan was taking the lead in railroad technology by building what would be the world's fastest train, Mitsubishi presented by far the lowest bid to provide new locomotives for towing on the Panama Canal, which was operated by the U.S. Army. Even with the higher price allowances ensured under the Buy American Act, the bid was lower than that of the next lowest bidder, an American company based in Ohio.

Nevertheless, the Army argued that they should reject it on the basis of national security. This case was considered especially significant as the first postwar instance of a Japanese company entering the American market for heavy industrial equipment and participating in a major government contract. The Japanese government lodged a formal protest against the Army's move, and Mitsubishi ultimately won the contract after the Office of Civil and Defense Mobilization determined that this would not threaten national security. But the effort to block their bid reflects growing anxieties about low-priced, high-quality foreign competition now facing American firms even for federal government contracts.[112]

Later the same year, a federal agency again stepped in to put a thumb on the scale for an American heavy industrial firm competing against a Japanese company for a government contract. Purchasing new hydraulic turbines for federal power plants, the Bureau of Reclamation created a new rule limiting the number of contracts that could go to a first-time provider, effectively limiting the Japanese bidder, Hitachi (which also produced components for the bullet train), to only one of the three contracts on which it had submitted the lowest bid.[113] This rule, however, could only provide temporary respite. Two years later, the *Wall Street Journal* reported that, for the third time in two weeks, a Japanese concern had won a U.S. Interior Department contract for turbines. Mitsubishi received a million-dollar contract, and Hitachi won two contracts, together worth a total of nearly four million dollars.[114] What was newsworthy was not that there was foreign competition—wrangling over lower-priced foreign competition in this industry had been going on for years—but that the companies presenting the winning bids were in Japan, which had previously purchased American heavy equipment without selling any to the United States.[115] Whether the reaction to a winning Japanese bid reflects a tinge of racism or simply shock at such a strong showing from a country not known as an exporter of heavy industrial goods, the point is that the success of Japanese companies in American government contracts presented an ambivalent image of Japan: as a country that could produce high-quality industrial products at a very low cost, it could be seen as both an equal and a threat.

Though the latter example does not directly involve railway equipment, by the time of the Satō-Johnson meetings, the bullet train may have affected American views of companies like Hitachi, which featured it prominently in their advertising as a symbol of their company's high quality and versatile production of advanced technologies, ranging from industrial machinery to consumer goods. In their advertisement in the New York World's Fair guide, Hitachi even claimed the train for themselves entirely, announcing, "Hitachi launches a new era in railroading with the Hitachi Superexpress, a train that streaks down the tracks at a phenomenal 155 mph."[116] The blurred train in the photograph conveys a sense of its unmatched speed, using its status as the world's fastest as a testament to all of the company's products (fig. 5). With advertisements like this, Japanese producers linked the bullet train—fast, powerful, efficient, and superior to any rail system in the United States—to the growing challenge of Japanese industry as a whole, suggesting that challenge stemmed not primarily from unfair trade practices, as American protectionists argued, but rather from the capacity for technological innovation, precision engineering, and flawless production.

The tensions over Japanese heavy industry firms winning U.S. government contracts were heightened just prior to the Satō-Johnson meeting by the prospect of federal support for mass transit construction and research into high-speed rail comparable to Japan's new bullet train. The Mass Transit Law, allocating federal funds for construction of mass transit infrastructure, passed on June 30, 1964, only after strong opposition was mollified by inclusion of the so-called Saylor Amendment requiring that all equipment procured under the new law be manufactured in the United States. This was much more stringent than existing policies, which merely gave American concerns a leg up in open competition for contracts, rather than barring foreign firms entirely.[117] Though Johnson, along with congressional Democrats, objected to the amendment, he saw no choice but to sign the bill as the only way to get mass transit legislation enacted over vehement Republican opposition.

The Japanese government immediately protested. On July 2, two days after the law was passed and a week before Johnson signed it, Japanese ambassador Takeuchi Ryūji told Secretary of State Dean Rusk and other top officials that passage of the bill would deal a severe blow to Japanese firms that had already received inquiries and entered into negotiations to provide equipment for various U.S.-based projects that would take advantage of federal funds. Beyond those specific cases, the Japanese government expressed concern that the idea behind the amendment would be extended

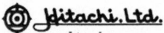

FIGURE 5. Hitachi's advertisement in the New York World's Fair guide showed off the bullet train. (Time-Life Books, 1964.)

to other fields.[118] That concern may have been intensified by an announcement just a few weeks before the Satō-Johnson meeting that the federal government would be requesting funds to study the implementation of high-speed rail transportation in the Northeast Corridor, followed by Johnson's promise to do so in his 1965 State of the Union address on January 4, in which he noted the necessity of modern transportation to economic growth.[119] As the administration expected, Satō expressed objections to the Mass Transit Law's Saylor Amendment rule in his meetings with Johnson, who responded with a promise that his administration was working for its repeal and complained to Satō of his daily confrontations with senators who "jump down his throat because of problems arising from Japanese imports."[120] Internally, American officials expressed doubt about their ability to make good on that promise, indicating the tightening hold of protectionist sentiments.[121]

Johnson did not share protectionists' fears of incursions by Japanese industry, not only because of his commitment to free trade but also because he was more focused on security issues, like nuclear proliferation, which for his administration was the most important technology issue for the bilateral meeting. At a dinner honoring Satō at the end of their first day of meetings, Johnson toasted the prime minister with a call for joint efforts "to make the world a better place through technology."[122] For Johnson, technological development was an area for cooperation with Japan, not competition, and his talks with Satō resulted in plans for bilateral programs in space exploration and medical science.[123] But in that approach, he fought against a trend in both industry and popular opinion that complicated Japan's technology diplomacy initiatives.

The push for a Northeast Corridor bullet train resulted in the introduction of the slower American-made Metroliner in 1969, and Japanese rail components did not become a significant topic in trade disputes. But the dynamic represented in the Satō-Johnson meetings—Japanese technological advances both promoting and impeding a smooth partnership—continued to characterize the U.S.-Japan relationship for decades after. The goals that inspired technology transfer to Japan after the late 1940s continued to influence American policy well into the Cold War era: in the ideological showdown between communism and capitalism, a strong Japanese economy ensured the continuity of conservative government and provided a showcase for the capitalist system. But in the mid-1960s, the very success of these policies shifted Japan's position vis-à-vis the United States and within the international community as a whole. The bullet train was one tool used to establish that position. For government entities (JNR, JETRO, and the Satō administration) the bullet train meant prowess in transportation infrastructure, contributing to Japan's emerging role as provider of technical assistance and symbolizing the "new Japan" as an industrial and technological power in world markets. This significance was challenged by those who perceived more of a threat than a promise in Japanese technology: Americans who had buried their fears of a strong Japan under a thin layer of cherry blossoms and industrialists and politicians concerned about a rising competitor. At every level, from individual to international, infrastructure exerted its enchanting power in varied ways, and the bullet train reverberated with ever-changing meanings. The New Tōkaidō Line is now over half a century old, but it continues to entrance, inspiring new dreams for new times.

Conclusion

Bullet Train Dreams in the Twenty-First Century

THE SUCCESSFUL DEBUT of "the world's fastest train" changed perceptions of what railways meant for Japan. Never just a fast and convenient means of transportation, trains had doubled as malleable symbols of the promise and threat of modernity since their nineteenth century arrival in Japan. In the context of the 1960s, the bullet train was wrapped up with contrasting views of Japan's past, present, and future; as a result, it was firmly imprinted on the national image. As Christopher Hood demonstrates, the expanded high-speed rail system, now extending across the entire archipelago, continues to function in varied ways as a "symbol of modern Japan."[1] Dubbed "zero-series" with the development of new models in the 1980s, the original train that is the focus of this book has held particular and changing meanings. Recent displays and representations of the zero-series show how the aesthetic power of the bullet train has outlasted its technical function, still making meaning years after its final journey.

The New Tōkaidō Line was built for a utilitarian purpose, but public celebrations of its development and construction made it into a monument to the successes of postwar Japan. Monuments are human landmarks that link the past to the future by concretizing a people's "collective force," the shared spirit or feelings of a particular time, into a form that outlives the moment of its creation to be inherited by future generations.[2]

215

An infrastructural construction project that is not built with an eye toward commemoration can be transformed into a monument through its public representations, which influence "encounters with the reality of the completed infrastructure and, by extension, with the affected landscape."[3] The monumentalization of the original bullet train began with JNR's promotion of the line under construction; for instance, the film discussed in the introduction to this book overlaid the process of planning and construction with a story of the triumph of humanity over nature through dedication and technological innovation. The preceding chapters have shown how other groups and individuals challenged, reinterpreted, or redirected this narrative. The official effort at monumentalization continued after the zero-series was decommissioned and, therefore, no longer involved in the mundane daily operations of the line. Recent representations by a successor to JNR, the privatized regional East Japan Railway Company (or JR East), deploy the zero-series (representing the New Tōkaidō Line) as a symbol of postwar Japan's economic growth and technological and industrial prowess. But like the promotional work done by JNR at the time of the line's development and debut, such posthumous treatment can slip away from those who seek to control it.

JR East's effort at monumentalization of the New Tōkaidō Line can be found in its Railway Museum, where the lead car of the Hikari Number One that made the inaugural run from Tokyo to Osaka in October 1964 is now ensconced in a room of its own. Here, JNR's original narrative about the bullet train echoes into the present day. The museum houses a display of retired locomotives and coaches that meanders through the technological and design history of Japan's railways, from the 1871 British-built locomotive that ran on Japan's first line to the earliest Japanese-built electric locomotive (manufactured in 1919), to a 1982 200-series bullet train, designed to deal with heavy snow as the network expanded northward. The biggest spotlight shines on the first bullet train. At the far end of the hall, marking the culmination of the historical journey, a zero-series cab stands near the 200-series model, surrounded by posters describing characteristics of each train and the high-speed rail system as a whole, such as technological and design innovations, passenger amenities, platform design, and safety equipment. A signboard is titled in Japanese, English, Chinese, and Korean, "Best of Advanced Railway Technologies: Start of

Tokaido Shinkansen."[4] The accompanying text notes the wartime anteced-
ents and technological advances needed to achieve the goal of world's fastest operating speed. A nearby display of bullet train–themed popular culture materials—including postcards, stamps, and special issues of popular
magazines marking its opening; cups and flatware decorated with its iconic
image; recordings of songs celebrating its speed; and of course numerous
toys, games, and books—evokes a feeling of popular awe and excitement.

In its separate room, the Hikari Number One (the only train car to receive such close and extensive attention in the museum) is surrounded
by text and photographs providing a detailed history of the development
and opening of the New Tōkaidō Line. Perhaps in recognition of the bullet
train's global impact and diplomatic significance, explanations here are bi-
lingual, with English translations accompanying most of the Japanese text.
Inside the car, interactive touch screens allow visitors to learn everything
they might want to know about the bullet train. Or, rather, everything that
JR East would like for them to know. The triumphant narrative of progress
and innovation that pervades the bullet train display and the museum as a
whole presents a popular story of modern Japanese history through trains
and stakes a claim about the importance of high-speed rail in contemporary
Japan. At the same time, the subsequent development of the system has rendered even this "train of the future" quaint. As one museum visitor recently
noted, given the introduction of sleeker and even more aerodynamic trains
over the decades, the feeling inspired by the "face" of the original train, with
its "dumpling" nose, changed from one of speed to something more "cute."[5]

Other voices contest the company's heroic portrayal, performing a kind
of counter-monumentalization by evoking a "collective force" that under-
mines the image of a united push for economic growth and industrial development. For instance, artist Kazama Sachiko uses nightmarish representations of fast trains to put a crack into that monumentalizing depiction.
In her early twenty-first-century works, the zero-series bullet train and one
of its wartime precursors, the SMR's Asia Express—highly recognizable objects from two historical moments—embody critiques of present-day Japanese society. Like the designers of the Railway Museum displays, Kazama
recognizes the aesthetic power of historic trains, but she harnesses that
power toward very different ends. Far from celebrating past events as steps
toward contemporary success, her works criticize the present by relating it

to an ugly past. Kazama's woodcut prints, combining social critique with an element of humor, anthropomorphize the trains, turning them into alien-type creatures or menacing robots. One of the earliest examples of this is her 2002 print *DANGAN-LESCHER* (*Dangan resshā*), one of a series titled *A Man for Remodeling the Japanese Archipelago* (*Rettō kaizō ningen*) (fig. 6). The print title, a near-homophone for "bullet train" (*dangan ressha*), is written with the character for "bullet" and the katakana *resshā*, which Kazama romanizes as "Lescher" for the English title. The series title is a reference to Tanaka Kakuei's 1972 book *A Plan for Remodeling the Japanese Archipelago* (*Nihon rettō kaizō ron*), which called for nationwide expansion of the bullet train system; in fact, the series includes a woodcut rendition of Tanaka's photograph that graces the cover of the book's original edition.[6] In this context, the book stands for the government's prioritization of economic growth and expansion of infrastructure, the consequences of which are critiqued, in this particular print, through Kazama's representation of the remodeled Japan as a monstrous creature: a zero-series train forms the head, back, and tail of a humanoid creature whose right arm ends in a cannonlike apparatus, suggesting that the remodeling work done by the bullet train has resulted in a grotesque militarized national body.

The bullet train plays a more innocent role as a victim of destruction in a 2007 print that also features the Asia Express: *A Steam Whistle, Mantetsu-man Manifest* (*Kiteki hitokoe, Mantetsujin arawaru*) (fig. 7). This title makes two historical references: "A steam whistle" is the opening phrase of a well-known Meiji-era railway song consisting of a verse for each station of the original Tōkaidō Main Line; and "Mantetsu" is the Japanese abbreviation of South Manchurian Railway, so "Mantetsu-man" (*Mantetsujin*) mimics the label many SMR employees had embraced for themselves. In this print, the locomotive of the old Asia Express forms the steam-spewing head of a giant robot, identifiable as an embodiment of the Asia Express not only by the distinctive shape of the locomotive/head but also by the SMR insignia adorning its midsection. Mantetsu-man stands astride a wrecked zero-series bullet train in Shiodome, the site of the Tōkaidō Main Line's first Tokyo Terminal, Shimbashi Station.[7] The damaged train's nose is shattered, a fallen utility pole has crashed into it from above, and smoke is pouring from the lead car. This print depicts the Asia Express and the bullet train in the opposite relationship from that posited in the 1960s nostalgia considered

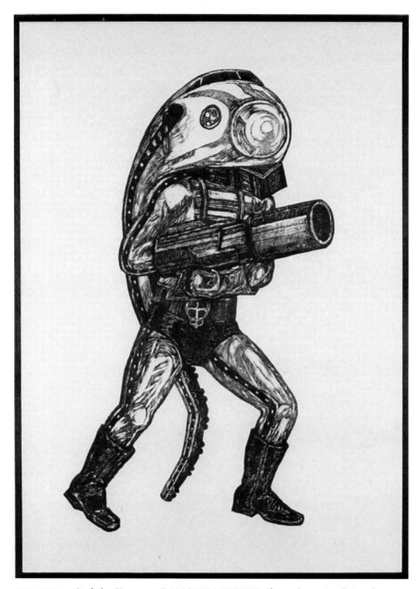

FIGURE 6. Sachiko Kazama, *DANGAN-LESCHER* (from the series "Man for remodeling the Japanese archipelago"), 2002, woodcut print (panel, Japanese paper, sumi ink). Photo: Keizo Kioku. © Sachiko Kazama. Courtesy of MUJIN-TO Production, Tokyo.

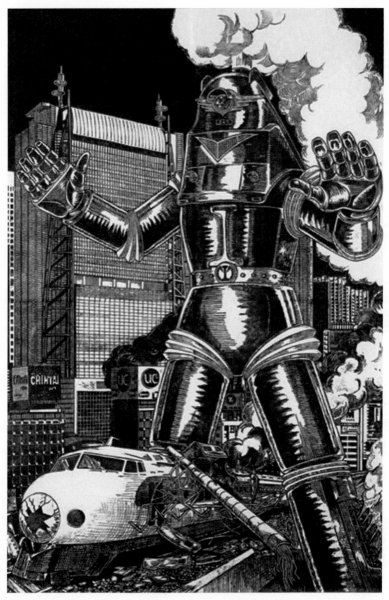

FIGURE 7. Sachiko Kazama, *A Steam Whistle, Mantetsu-man Manifest*, 2007, woodcut print (panel, Japanese paper, sumi ink). Photo: Keizo Kioku. © Sachiko Kazama. Courtesy of MUJIN-TO Production, Tokyo.

in Chapter 4. Instead of a precursor to the bullet train—the necessary pre-condition for postwar prosperity—this embodied symbol of Japan's war-time imperialism is clearly the enemy of postwar success. Forgetting the violence of war and empire, Kazama's work warns, presents a dire threat to the peace and prosperity of postwar Japan. In a sense, Kazama's print performs a function similar to that of 1960s nostalgia for the Asia Express; it reconnects two historical moments that have been separated by radical change. However, Kazama's aim is not to help people forget the wartime past but rather to entreat them to remember it. Interviewed in 2014, the art-ist argued that symbols of industrial and economic growth obscure the his-tory of the Asia-Pacific War from public awareness, leading to a "condition of postwar dementia."[8] Her goal is to bring that past—not airbrushed mem-ories but the realities of oppression, militarism, and violence—back before the eyes of an amnesiac Japanese public.

The Railway Museum displays and Kazama's prints are just two exam-ples of the continued use of the historical bullet train as a key element of dreams (or nightmares) about contemporary Japan and its future. As Ka-zama's critique shows, visions of the future that use the bullet train as a symbol are not necessarily related directly to high-speed rail. Similarly, the trends described in the preceding chapters did not depend on the existence of a particular railway assemblage. The city of Kyoto was already promot-ing a modern, international identity, and some residents were becoming politically active in connection with the national movement against revi-sion of the U.S.-Japan Security Treaty, labor protests, and various local is-sues. Many factors, including other transportation and communications infrastructures, were integrating the Tōkaidō region and pushing the Japa-nese economy and society toward informationization. And numerous sites were used to reconstruct both public memory of the war years and Japan's image abroad. But the bullet train gave those developments a particular an-gle, and the fact that so many people took it up as an emblem of either the strengths or the failings of Japanese society imprinted the resulting charac-teristics of contemporary Japan with the image of the bullet train, helping to turn it into a national symbol. The bullet train is a useful vantage point from which to view its times because it was interpreted from many view-points through the concerns of the day, and through those multiple inter-pretations, it contributed to the dominant political, economic, and cultural characteristics of postwar Japan. Viewing the long 1960s through the bullet train window, therefore, accentuates the multiplicity of experiences of the era of high-speed economic growth and the contingency of its path. In the same way, considering contemporary legacies of those historical threads provides insight on present-day Japan.

RECURRENT DREAMS OF A SUPER-EXPRESS

This book has examined several different dreams of the super-express. As local and national government leaders and urban planners negotiated the contours of the line, intellectuals theorized about its economic and social impact, often contradicting official predictions about its effects. The general public learned about the new system and pondered its significance for themselves, for their city, and for their nation's past, present, and future. Some took action to attract a station or protest the line's proximity to their homes. Eventually, passengers rode the trains, and many wrote about their experiences. Outside Japan, reporters, railway specialists, policy makers, and travelers considered the meaning of the train in terms of their own countries' railroad systems and relations with Japan. Each of these stories continues to reverberate in more current dreams that reimagine society and the future through high-speed rail.

The planning of the New Tōkaidō Line inspired opposing efforts to influence its route and other aspects of construction. Now the prospects of yet another "world's fastest train"—the superconducting maglev line planned to run between Tokyo and Osaka at over 500 kilometers per hour—is inspiring present-day echoes of the 1960s in both efforts to attract a station and resistance against it. In their push for a Kyoto stop on the maglev line, officials created a multimedia public relations campaign using similar methods and rhetoric as those demonstrated by the city administration six decades ago. Beginning in 2014, the city produced a host of content—including a poster, pamphlet, video, and social media pages—harnessing the same ideas the Kyoto City Assembly used to convince JNR to bring the New Tōkaidō Line to Kyoto (see fig. 8). On the pamphlet cover (also produced as a poster), for instance, the futuristic train runs across the center of the page under the slogan "For Japan's future, [bring] the maglev (*rinia*) to Kyoto."[9] It is flanked by silhouettes of the most recognizable symbols of Kyoto's cultural significance: fans, temple buildings, and a large *torii* gate,

FIGURE 8. Cover of a digital pamphlet created by Kyoto City as part of their campaign to attract the maglev line, c. 2014.

with stylized pink cherry blossoms scattered all around. This image deploys the same metropolitan identity as 1960s claims that, because of its status as the storehouse of traditional culture, Kyoto is essential to a national project such as these high-speed railway lines. The city presents its call to "prepare the best route" for the maglev as being not for the benefit of Kyoto alone but for the future of the nation as a whole.

Current Kyoto mayor Kadokawa Daisaku also sounded familiar notes, though with a few updated terms. Discussing the request that the maglev plan be changed to include the city, he emphasized its national significance in terms of culture and tourism. Fifty years after the campaign that brought the zero-series through Kyoto, Kadokawa stated, "Kyoto's 'soft power' resources, such as history and culture, can play an important role in Japan's future development." Citing tourism, he asserted, "If the maglev bullet train does not come through Kyoto, the city will be displaced from Japan's national axis, which could hurt the national interest."[10] He added the claim that including Kyoto on the line would increase the opportunity for people to experience the historical charm of Japan that does not exist in Tokyo. Calling the city's attractions its soft power resources puts a contemporary label on the same arguments that his predecessors used to convince JNR that the bullet train needed Kyoto as much as the city needed a stop on the line.

Kyoto has not been able to replicate its previous success in changing the planned route. In contrast, Shizuoka Prefecture has effectively halted progress on the line by refusing to grant permission for construction, based on environmental concerns. Like the farmers fictionalized in an episode of the television show *Bullet Train* (*Dangan ressha*, discussed in Chapter 2), Shizuoka's cultivators of tea and oranges feared the impact of the new line's tunnels on the water resources essential to their fields. The concern of today's farmers is that underground water will run into a tunnel built for the new line, thus reducing a resource essential to their livelihoods, precisely the scenario of the 1964 television episode. In this case, too, the maglev project inspired similar reactions as the New Tōkaidō Line did decades before, but with a very different outcome. The difference may be related to the specific prefectural interests of Shizuoka, which stands to gain little from the new line: a maglev station is not planned for the prefecture, so it would suffer damages without reaping benefits.[11] But Shizuoka's ability to use environmental regulations to stop the project reflects changes in Japanese society in the intervening decades, including greater attention to the environmental risks of large infrastructural projects and other consequences of the prioritization of economic development.

Another significant factor in Shizuoka's ability to halt maglev construction is the radical expansion of the information society that the bullet train helped to usher in. The internet had already brought profound changes to economic structures and business operations that made futurologists place the bullet train in the middle of the emerging socioeconomic structures. Now, as the maglev's future hangs in the balance, the extensive telecommunications infrastructures developed in recent years support the social media that facilitate organization and help amplify protest. In addition, of course, the impact of the COVID-19 pandemic on business practices has brought doubt to estimations of ridership, weakening the very reasons for its existence.[12]

In the midst of excitement about new railway technologies and ever-increasing speeds, as Kazama's robot-train demonstrates, the memory work of previous fast trains continues. Four decades after its demise and a decade after the Asia Express boom of the 1960s had faded, the defunct SMR train enjoyed yet another afterlife: a powerful tool of imperial control was reincarnated as a means of reconciliation. In the early 1980s, a joint Sino-Japanese group of railroad enthusiasts joined hands with Chinese National Railways in an effort to use the train's aesthetic function—its power to elicit strong emotional responses from those who remembered it—to improve bilateral relations. The project began when a Japanese film crew in northeast China making a documentary about Chinese steam locomotives spied a train car with an unmistakable streamlined silhouette sitting on a track near where they were filming. Taking a closer look, they found the SMR insignia and the year 1934 etched into the chassis and knew it had to be the iconic train's first-class observation car. Soon after, a "Pashina" locomotive that had pulled the train was also located, inoperative but still intact, in a distant railyard. In response to interest from Japanese railroading circles, Chinese railway authorities had the locomotive moved to their railyard in Shenyang for restoration. In this case, a train with little or no technical function nevertheless exerted a powerful aesthetic force. Unable to run on its own power, the locomotive had to be pulled to Shenyang by another engine. Even so, a former *Yomiuri* correspondent, moved by the sight of the completely inert vehicle, described it as being "alive." Though repainted in unfamiliar colors, "the beautiful streamlined form bathed in the evening sun was truly the past itself."[13]

Two years later, a group of twenty-one Japanese rail enthusiasts traveled to Dalian for a train ride without a destination: the partially restored locomotive, operating at about a third of its former strength, pulled a single train car back and forth over a short track.[14] Like some of the bullet train tourism promoted in 1964, the ride itself was the object of the trip. Chinese rail authorities began to organize a longer nostalgic ride, in cooperation with the private Sino-Japanese Railroad Exchange Society, inviting former SMR employees and other Japanese residents of occupied Manchuria to enjoy a "Sino-Japanese Amity Test Run," a ten-hour ride from Dalian to Shenyang on a train pulled by the old Pashina locomotive. Promoters of this railroad diplomacy counted on the emotional impact of seeing the old locomotive and riding through once-familiar countryside—both part of the train's aesthetic function—for the success of the endeavor. Chinese organizers hoped that letting SMR employees once again experience the Asia Express of their memories would cultivate friendly feelings toward China, but the full-length journey never materialized.[15]

Excitement in Japan about the locomotive's resuscitation (the restoration was invariably described in terms of the train coming back to life) brought back several of the themes heard in the 1930s and again in the 1960s. Recounting of its "epochal" features accompanied comments that it had been designed and built by Japanese engineers in less than a year, thanks to the intense hard work and devotion of the SMR engineering team, praised (as in the past) for working night and day without rest. But the Chinese role in the joint effort had some effect in changing views of the train's new incarnation, as reports of the restoration noted that it was Chinese National Railways that brought the locomotive to their railyard and restored it to working order.[16]

Transportation infrastructure was a key attraction in the late twentieth-century Japanese nostalgia industry. Former colonizers yearned not for Manchuria as a place but rather for a full range of experiences associated with the Japanese empire in northeast China.[17] Travel on the Asia Express was an important part of that experience for at least the upper class of Japanese colonizers. Mariko Asano Tamanoi sees this memory industry as yet another mode of forgetting, "assisting the Japanese people to forget the power of their own state, which once dominated ordinary Chinese people in a place where they now entertain themselves."[18] The desire to attract Japanese tourists resulted in the restoration of a second Pashina locomotive that was found in August 2001. The following year, construction started on the Shenyang Steam Locomotive Museum, a prominent attraction of which was to be an operating Pashina locomotive, which would pull tourists around a 30-kilometer track. That particular plan did not come to fruition, but both locomotives, as well as the observation car, are now on display there in an exhibit that was opened to the public in May 2019 in the hopes that they would become a tourist draw for Japanese railway fans.[19] Viewed from the perspective of Japan's current tensions with China over its imperialist past, it is surprising that such concerns were set aside or silenced in order to remake the train as a tourist attraction. But lingering discomfort with its history is visible in the failure of various plans to operate the train, as well as in delays in putting it on display, as Chinese government authorities investigated whether its exhibition would be appropriate to policies on patriotic education.[20]

In Japan's international relations more broadly, the factors that created an opportunity for using infrastructure as a source of soft power in the 1960s are still relevant. Aid and trade remain at the center of Japanese foreign policy today, and popular foreign views of Japan still tend to center on its real and imagined technology. In recent years, China has followed a similar playbook, using impressive infrastructure for global events like the 2008 Beijing Olympics and the 2010 Shanghai Expo, as well as the multiple development projects of its Belt and Road Initiative, to revamp its global image. China is also vying with Japan for the informal title of world leader in high-speed rail through both exports and construction of a vast, growing, and accelerating domestic network. These efforts have prompted a similarly ambivalent global response. Scholars point out that the Orientalism of past eras has been replaced by a techno-Orientalism, in which Western views of Japan (and China) in terms of high technology—promoted and reinforced by copious self-representations—have a dehumanizing effect.[21] By presenting their nation in terms of technological objects like the bullet train, past Japanese image makers like JETRO's World's Fair planners helped disseminate technology-centered images of Japan. Today's leaders perpetuate that perception, for instance, by dressing up the prime minister as a Nintendo video game character for the closing ceremony of the 2016 Rio Olympics in order to invite the world to the 2020 Olympiad in Tokyo.

The bullet train's international reputation for successfully combining speed, safety, and efficiency seems to make it an excellent candidate for such national branding work. Indeed, the Japanese government actively promotes the export of high-speed rail and other infrastructural systems as an important element of its economic plans.[22] But on a popular level, the bullet train sometimes ends up reinforcing stereotypes, as news of minor incidents or details of operation explodes across the international media, in part because they accord with expectations about Japan. For instance, JR West's 2018 apology to its customers for a train's twenty-five-second premature departure became international news, in spite of the story's utter lack of newsworthiness, because it confirms clichés about Japanese precision and satisfies the current Western appetite for stories about "weird Japan."[23] But in this case, too, stereotypes were reinforced from within. What foreign news outlets called an "obsession with punctuality" is the regular operating standard that led railway management to feel that an apology was warranted.[24] And the impression of precision, a key element of quality and reliability, is itself an important selling point for Japanese infrastructure and technology exports.

The potentials and failings of monumentalizing infrastructure are visible in these recurrent dreams of high-speed rail. While Kyoto's leaders see the maglev as a new national monument that should not, therefore, bypass the nation's cultural heart, Shizuoka's leaders highlight its destructive force, reminding us that monuments also damage the environments in which they are built. Such instances of resistance take a chisel to top-down efforts at monumentalization. As a lasting pandemic has forced so much of work life into virtual spaces, the utility of ever-faster railways is being questioned. If Japanese society is changing rapidly, is high-speed rail still the right infrastructure to embody its collective force? The Chinese government's hesitancy to deploy the Asia Express as a soft power resource in bilateral relations reflects a desire to avoid creating a monument to Japanese imperialism in China. And efforts to export Japan's bullet train system are seeing little success, suggesting that the bullet train as a monument to Japanese technological innovation and railway prowess holds dwindling power. These examples show that the monumentalization of infrastructure is just as contested as the process of building it. A given technical function of infrastructure may be realized, but whether it has the power to hold meaning (and what meaning it holds) is up for grabs. Monumentalization may preserve a place for infrastructure in the popular imagination; however, the process that transforms infrastructure into a monument is never predetermined but rather the contingent result of the push and pull of competing dreams.

Notes

INTRODUCTION

1. For an overview of recent literature on this theme, see Hannah Appel, Nikhil Anand, and Akhil Gupta, "Introduction: Temporality, Politics, and the Promise of Infrastructure," in Nikhil Anand, Akhil Gupta, and Hannah Appel, eds., *The Promise of Infrastructure* (Durham, NC: Duke University Press, 2018), 1–28.

2. Penny Harvey and Hannah Knox, "Enchantments of Infrastructure," *Mobilities* 7, no. 4 (November 2012): 521–536.

3. Brian Larkin, "The Politics and Poetics of Infrastructure," *Annual Review of Anthropology* 42 (August 2013): 333. Also see Brian Larkin, "Promising Forms: The Political Aesthetics of Infrastructure," in Nikhil Anand, Akhil Gupta, and Hannah Appel, eds., *The Promise of Infrastructure* (Durham, NC: Duke University Press, 2018), 175–202.

4. John Durham Peters, *The Marvelous Clouds: Toward a Philosophy of Elemental Media* (Chicago: University of Chicago Press, 2015), 35–36.

5. Michael Adas, *Machines as the Measure of Men: Science, Technology, and Ideologies of Western Dominance* (Ithaca, NY: Cornell University Press, 1989); Michael Adas, *Dominance by Design: Technological Imperatives and America's Civilizing Mission* (Cambridge, MA: Harvard University Press, 2009).

6. On the technical development of the bullet train, see Tōru Koyama, "The Shinkansen (Bullet Train): A New Era in Railway Technology," in Shigeru Nakayama and Kunio Gotō, eds., *A Social History of Science and Technology in Contemporary Japan*, vol. 3 (Melbourne: Trans Pacific Press, 2006), 379–389; and Takashi Nishiyama, *Engineering War and Peace in Modern Japan, 1868–1964* (Baltimore: Johns Hopkins University Press, 2014).

7. Harvey and Knox, "Enchantments of Infrastructure," 525–528.

8. Wolfgang Schivelbusch, *The Railway Journey: The Industrialization of Time and Space in the 19th Century* (Berkeley: University of California Press, 1986), 23; Leo Marx, "The Impact of the Railroad on the American Imagination, as a Possible Comparison for the Space Impact," in Bruce Mazlish, ed., *The Railroad and the Space Program: An Exploration in Historical Analogy* (Cambridge, MA: MIT Press, 1965), 202–216; Steven J. Ericson, *The Sound of the Whistle: Railroads and the State in Meiji Japan* (Cambridge, MA: Harvard University Asia Center, 1996), 62–63.

9. Yoshihisa Tak Matsusaka, *The Making of Japanese Manchuria, 1904–1932* (Cambridge, MA: Harvard University Asia Center, 2001), 147–148, 394.

10. Railways were part of a broad effort to rebuild relationships with other Asian nations while bolstering Japan's national economy through development assistance. On that context, see Hiromi Mizuno, "Introduction: A Kula Ring for the Flying Geese: Japan's Technology Aid and Postwar Asia," in Hiromi Mizuno, Aaron S. Moore, and John DiMoia, eds., *Engineering Asia: Technology, Colonial Development and the Cold War Order* (London: Bloomsbury, 2018), 1–40.

11. Andrew Gordon, "Managing the Japanese Household: The New Life Movement in Postwar Japan," *Social Politics* 4, no. 2 (1997): 247.

12. The Annual Economic White Paper (*Nenji keizai hakusho*) for 1956 is at https://www5.cao.go.jp/keizai3/keizaiwp/wp-je56/wp-je56-00001i.html; the famous assertion that "the postwar is over" (*mohaya sengo de wa nai*) is in the conclusion at https://www5.cao.go.jp/keizai3/keizaiwp/wp-je56/wp-je56-010501.html.

13. Langdon Winner, "Do Artifacts Have Politics?" *Daedalus* 109, no. 1 (Winter 1980): 127–128.

14. Dennis Rodgers and Bruce O'Neill, "Infrastructural Violence: Introduction to the Special Issue," *Ethnography* 13, no. 4 (2012): 404.

15. Christopher P. Hood, *Shinkansen: From Bullet Train to Symbol of Modern Japan* (New York: Routledge, 2006); Nishiyama, *Engineering War and Peace*; Fujii Satoshi, *Shinkansen to nashonarizumu* (Tokyo: Asahi Shinbun Shuppan, 2013); Kondō Masataka, *Shinkansen to Nihon no hanseiki* (Tokyo: Kōtsū Shinbunsha, 2010). The many examples of books by former JNR employees include Suda Hiroshi, *Tōkaidō Shinkansen sanjū-nen* (Tokyo: Taishō Shuppan, 1994); and Yamanouchi Shūichirō, *Shinkansen ga nakattara* (Tokyo: Asahi Shinbunsha, 2004).

16. For example, Michael Fisch, *An Anthropology of the Machine: Tokyo's Commuter Train Network* (Chicago: University of Chicago Press, 2018); Alisa Freedman, *Tokyo in Transit: Japanese Culture on the Rails and Road* (Stanford, CA: Stanford University Press, 2011); Andrew Gordon, *Fabricating Consumers: The Sewing Machine in Modern Japan* (Berkeley: University of California Press, 2011); Barak Kushner, *Slurp! A Social and Culinary History of Ramen—Japan's Favorite Noodle Soup* (Boston: Global Oriental, 2012); Ian J. Miller, *The Nature of the Beasts: Empire and Exhibition at the Tokyo Imperial Zoo* (Berkeley: University of California Press, 2013); Kerry Ross, *Photography for Everyone: The Cultural Lives of Cameras and Consumers in Early Twentieth-Century Japan* (Stanford, CA: Stanford University

Press, 2015); and George Solt, *The Untold History of Ramen: How Political Crisis in Japan Spawned a Global Food Craze* (Berkeley: University of California Press, 2014).

17. Edward Said, *Orientalism* (New York: Pantheon Books, 1978); David Morley and Kevin Robbins, *Spaces of Identity: Global Media, Electronic Landscapes and Cultural Boundaries* (New York: Routledge, 1995), 6, 147–173; David S. Roh, Betsy Huang, and Greta A. Niu, *Techno-Orientalism: Imagining Asia in Speculative Fiction, History, and Media* (New Brunswick, NJ: Rutgers University Press, 2015); Toshiya Ueno, "Japanimation and Techno-Orientalism," in Bruce Grenville, ed., *The Uncanny: Experiments in Cyborg Culture* (Vancouver: Arsenal Pulp Press, 2002), 223–236.

18. On early railroad development in Japan, see Ericson, *Sound of the Whistle*.

19. These numbers were ubiquitous during the years of planning, construction, and early operation of the new line; see, for instance, JNR Chief Engineer Fujii Matsutarō's foreword to a special issue on the New Tōkaidō Line in *Japanese Railway Engineering* 5, no. 4 (December 1964): 5.

20. Ericson, *Sound of the Whistle*, 274–275.

21. Ryuichi Hamaguchi, "The Tokaido New Trunk Line," part 1, *Japan Architect*, no. 110 (July 1965): 23–25.

22. "Shinkansen are kore," *Dōboku gakkai shi* 49, no. 10 (1964): 47.

23. Casper Bruun Jensen and Atsuro Morita, "Infrastructures as Ontological Experiments," *Engaging Science, Technology, and Society* 1 (2015): 81–87.

24. Hiraku Shimoda makes this argument through analysis of an episode of the popular documentary television series *Project X* that aired on May 9, 2000: Hiraku Shimoda, "'The Super-Express of Our Dreams' and Other Mythologies About Postwar Japan," in Benjamin Fraser and Steven D. Spalding, eds., *Trains, Culture, and Mobility: Riding the Rails* (Lanham, MD: Lexington Books, 2012), 264–265.

25. This narrative can be found in materials produced by JNR, such as *Railway Technical Research Institute* (Tokyo: JNR, 1962); *Shinkansen '62* ([Tokyo]: Nihon Kokuyū Tetsudō, 1962); *Shinkansen o ichi nichi mo hayaku* ([Tokyo]: Nihon Kokuyū Tetsudō, [1963]); *Tōkaidō Shinkansen* ([Tokyo]: Nihon Kokuyū Tetsudō, [1964]); and a documentary film commissioned by JNR, Koyama Seiji, dir., *Tōkaidō Shinkansen* (Shin Riken Eiga, 1964).

26. "Shinkansen no shōkai eiga jōei," *Asahi shinbun*, April 28, 1965, 10.

27. Koyama, *Tōkaidō Shinkansen*.

28. Koyama, *Tōkaidō Shinkansen*.

29. Christina Klein, *Cold War Orientalism: Asia in the Middlebrow Imagination, 1945–1961* (Berkeley: University of California Press, 2003), 6–7.

30. Freedman, *Tokyo in Transit*, 1.

31. Marian Aguiar, *Tracking Modernity: India's Railway and the Culture of Mobility* (Minneapolis: University of Minnesota Press, 2011), xii.

32. Examples that will be examined in the following chapters include Nishiguchi Katsumi, *Shinkansen* (Tokyo: Kōbundō, 1966); Shimada Kazuo and Yokomitsu

Akira, *Dangan ressha: NHK karā eiga*, directed by Kishida Toshihiko (Nippon Hōsō Kyōkai, 1964), scripts held at the Tsubouchi Memorial Theatre Museum Library, Waseda University; Satō Junya, dir., *Shinkansen daibakuha* (Tōei, 1975); Kajiyama Toshiyuki, *Yume no chōtokkyū: Chōhen suiri shōsetsu* (Tokyo: Kōbunsha, 1963); and Masumura Yasuzō, dir., *Kuro no chōtokkyū* (Daiei, 1964).

CHAPTER 1

1. Nishiguchi Katsumi, *Shinkansen* (Tokyo: Kōbundō, 1966), 6–22.

2. The most well-known case is that of Gifu-Hashima Station, which is discussed in Chapter 2 and in Christopher P. Hood, *Shinkansen: From Bullet Train to Symbol of Modern Japan* (New York: Routledge, 2006), 72–74.

3. Walter Benjamin, "Theses on the Philosophy of History," in Hannah Arendt, ed., *Illuminations*, trans. Harry Zohn (New York: Schocken Books, 1968), 257; Ann Laura Stoler, *Along the Archival Grain: Epistemic Anxieties and Colonial Common Sense* (Princeton, NJ: Princeton University Press, 2008).

4. Kyoto Shikai Jimukyoku Chōsaka, ed., *Kyoto shikai junpō* (hereafter *KSJ*).

5. Amakawa Akira, "The Making of the Postwar Local Government System," in Robert E. War and Sakamoto Yoshikazu, eds., *Democratizing Japan: The Allied Occupation* (Honolulu: University of Hawai'i Press, 1987), 251–282.

6. On anti-Anpo protests as grassroots efforts to shape Japan's democracy, see Wesley Sasaki-Uemura, *Organizing the Spontaneous: Citizen Protest in Postwar Japan* (Honolulu: University of Hawai'i Press, 2001).

7. For instance, Nikhil Anand, *Hydraulic City: Water and the Infrastructures of Citizenship in Mumbai* (Durham, NC: Duke University Press, 2017); Leo Coleman, *A Moral Technology: Electrification as Political Ritual in New Delhi* (Ithaca, NY: Cornell University Press, 2017); Antina von Schnitzler, *Democracy's Infrastructure: Techno-Politics and Protest after Apartheid* (Princeton, NJ: Princeton University Press, 2016).

8. Ellis S. Krauss, "Opposition in Power: The Development and Maintenance of Leftist Government in Kyoto Prefecture," in Ellis S. Krauss and Scott E. Flanagan, eds., *Political Opposition and Local Politics in Japan* (Princeton, NJ: Princeton University Press, 1981), 383–424.

9. "Kyoto–Tokyo kaisoku '2 jikan 40 fun,'" *Kyoto shinbun* August 19, 1964, 3.

10. Jeffrey E. Hanes, *The City as Subject: Seki Hajime and the Reinvention of Modern Osaka* (Berkeley: University of California Press, 2002), 174–193.

11. The contested process of identity formation through urban planning resembles the case of Hiroshima examined by Ran Zwigenberg, though he asks not how individuals resisted the top-down narrative thus created, but rather why some embraced it (Ran Zwigenberg, *Hiroshima: The Origins of Global Memory Culture* (New York: Cambridge University Press, 2014), 23–93).

12. On the technical and aesthetic functions of infrastructure, see Brian Larkin, "The Politics and Poetics of Infrastructure," *Annual Review of Anthropology* 42 (August 2013): 327–343.

13. Similar dynamics occur in any case of major infrastructure planning, such as dams and nuclear power stations. See Eric G. Dinmore, "'Mountain Dream' or the 'Submergence of Fine Scenery'? Japanese Contestations over the Kurobe Number Four Dam, 1920–1970," *Water History* 6 (2014): 315–340; Martin Dusinberre, *Hard Times in the Hometown: A History of Community Survival in Modern Japan* (Honolulu: University of Hawai'i Press, 2012), 150–188.

14. "Shin Tōkaidō sen kōsu kettei," *Kyoto shinbun*, November 15, 1959, 1; *KSJ* (1959), no. 371, p. 435; *KSJ* (1959), "Shōwa 34-nen kaiko," pp. 100–101. The two straight-line routes considered crossed the Suzuka Mountain Range at the Suzuka Pass and the Happū Pass on the border of Mie and Shiga Prefectures.

15. *KSJ* (1959), no. 370, p. 427; *KSJ* (1959), "Shōwa 34-nen kaiko," p. 102.

16. *KSJ* (1958), no. 323, pp. 20–21.

17. *KSJ* (1958), no. 345, p. 17.

18. *KSJ* (1958), no. 347, p. 18.

19. *KSJ* (1959), no. 370, pp. 426–427.

20. *KSJ* (1959), no. 372, pp. 476–477.

21. *KSJ* (1959), no. 372, p. 477.

22. These efforts are noted and summarized throughout the city assembly records (*KSJ*) for the years 1958 to 1964.

23. Alice Y. Tseng, *Modern Kyoto: Building for Ceremony and Commemoration, 1868–1940* (Honolulu: University of Hawai'i Press, 2018), 215.

24. Tseng, *Modern Kyoto*, 220.

25. Isomura Eiichi, "Toshika no shinten to sono eikyō: Nishi Nihon kaihatsu no kadai," *Chiiki kaihatsu* 7, no. 58 (July 1969) 33; Isomura Eiichi, "Shinkansen to tochi kaihatsu riron," *Chiikigaku kenkyū*, no. 3 (September 1973): 49.

26. *KSJ* (1958), no. 345, p. 18.

27. "Shōwa 34-nen no shikai o kaerimite," *KSJ* (1959), "Shōwa 34-nen kaiko," [n.p.].

28. "Kyoto Kokusai bunka kankō toshi kensetsu hō" (1950). The text of this and similar laws directed toward other cities is available through *e-Gov: Denshi seifu no sōgō madoguchi*, https://www.e-gov.go.jp.

29. "Kyoto shimin kenshō," *Kyōtoshi jōhōkan*, https://www.city.kyoto.lg.jp/sogo/page/0000184650.html.

30. Yoneyama Toshinao, "Bunka kankō toshi," in Hayashi Tatsusaburō, ed., *Kyoto no rekishi*, vol. 9 (Tokyo: Gakugei shorin, 1976), 511.

31. "Shōwa 34-nen no shikai o kaerimite," *KSJ* (1959), "Shōwa 34-nen kaiko," [n.p.].

32. "Kyō no 'kokusai toshidzukuri,'" *Kyoto shinbun*, November 25, 1959, 11.

33. "Shōwa 37-nen no shikai o kaerimite," *KSJ* (1962), "Shōwa 37-nen kaiko," [n.p.].

34. *KSJ* (1959), no. 372, pp. 476–477.

35. "Tōkaidō shinsen wa junchō," *Kyoto shinbun*, January 14, 1960, evening ed., 7.

36. *KSJ* (1960), no. 399, p. 436.
37. *KSJ* (1960), no. 401, pp. 465–466.
38. *KSJ* (1960), no. 389, pp. 72–73; *KSJ* (1964), no. 553, p. 514.
39. *KSJ* (1964), no. 552, p. 496.
40. *KSJ* (1964), no. 553, p. 513.
41. "Shi to Kokutetsu ga kyōryoku," *Kyoto shinbun*, August 19, 1964, city ed., 12.
42. "Kyoto–Tokyo kaisoku '2 jikan 40 fun.'"
43. "Kamogawa tekkyō no kōji saikai," *Kyoto shinbun*, October 26, 1961; *KSJ* (1961), "Kokutetsu taisaku mondai ni tsuite (1-gatsu—12-gatsu)," pp. 129, 131, 137.
44. Krauss, "Opposition in Power," 399–400.
45. "Kyoto–Tokyo kaisoku '2 jikan 40 fun.'"
46. *KSJ* (1960), no. 389, pp. 72–73.
47. *KSJ* (1960), no. 389, pp. 72–73.
48. "Chōtokkyū Kyoto teisha ni kyōryoku," *Kyoto shinbun*, August 15, 1964, 2.
49. "Iitai hōdan," *Kyoto shinbun*, September 22, 1964, evening ed., 3.
50. Iwai Seiji (Kyoto Chamber of Commerce and Industry vice president), Sagawa Kazuo (Kyoto Prefectural Assembly chair), and Taoka Ryōichi (Kyoto University emeritus professor) in "Kyoto teisha jimoto yorokobi no koe," *Kyoto shinbun*, August 19, 1964, 3.
51. "'Hikari' wa hatten no shinboru desu," *Kyoto shinbun*, September 29, 1964, 7.
52. "'Hikari' wa hatten no shinboru desu," 7.
53. "Iitai hōdan."
54. On the view of modern infrastructure as being new, clean, and hard, see Rudolf Mrázek, *Engineers of Happy Land: Technology and Nationalism in a Colony* (Princeton, NJ: Princeton University Press, 2002), 8.
55. "Kindaika e biru būmu," *Kyoto shinbun*, August 18, 1964, city ed., 15.
56. "Kindaika e biru būmu." The word "Kyoto," written in katakana in the article, is rendered in all capital letters here to capture that emphasis.
57. "Tetsudō yomoyamabanashi," *Kyoto shinbun*, September 26, 1964, evening ed., 3.
58. "Koto rashii yosooi," *Kyoto shinbun*, September 27, 1964, morning ed., 14.
59. Penny Harvey and Hannah Knox, "Enchantments of Infrastructure," *Mobilities* 7, no. 4 (November 2012): 529.
60. Sasaki-Uemura, *Organizing the Spontaneous*, 3.
61. Sasaki-Uemura, *Organizing the Spontaneous*, 6.
62. Sasaki-Uemura, *Organizing the Spontaneous*, 27–34.
63. "Tōkaidō shinsen wa junchō."
64. "Shinkansen rūto kaenu," *Kyoto shinbun*, March 12, 1960, Daini Shiga ed., 13.
65. For instance, *KSJ* (1959), no. 385, p. 842.
66. "Tochi shūyōhō o tekiyō e," *Yomiuri shinbun*, Kyoto ed., January 14, 1960, 10. At the time, Yamashina was part of Higashiyama Ward.

67. The example of the Meishin Expressway came up explicitly during a city assembly discussion of opposition from residents: *KSJ* (1960), no. 395, p. 288.

68. *KSJ* (1960), no. 394, p. 240; *KSJ* (1960), no. 395, pp. 287–291; "Shin Tōkaidō sen Kyoto kōsu rokugatsu ni zen rosen kakutei," *Kyoto shinbun*, March 10, 1960, Daini Shiga ed., 13.

69. "Kōkyō yōchi tokushu ni chōsakai," *Yomiuri shinbun*, July 31, 1960, evening ed., 1.

70. "Kōkyō yōchi tokushuhō kyō shikō," *Yomiuri shinbun*, August 17, 1961, morning ed., 2.

71. "Shinkansen nado sanken o 'tokutei' ni," *Yomiuri shinbun*, October 20, 1962, morning ed., 2.

72. *KSJ* (1959), no. 385, p. 842.

73. *KSJ* (1960), no. 394, p. 239

74. *KSJ* (1960), no. 389, pp. 72–73.

75. *KSJ* (1960), no. 397, p. 49.

76. *KSJ* (1960), no. 398, p. 413.

77. *KSJ* (1960), no. 394, pp. 238–239.

78. Nishiguchi, *Shinkansen* (1966), 21–22.

79. Examples include Nishiguchi, *Shinkansen* (1966), 32–33, 45, 55, 94, 136.

80. For instance, *KSJ* (1959), no. 370, pp. 429–430; *KSJ* (1959), no. 385, p. 841.

81. "Tōkaidō shinsen wa junchō." Also see "Shinkansen rūto kaenu."

82. *KSJ* (1961), "Shōwa 36-nen o kaiko shite," p. 132; *KSJ* (1962), "Shōwa 37-nen o kaiko shite," p. 136.

83. For example, Nishiguchi, *Shinkansen* (1966), 40, 55.

84. *KSJ* (1959), no. 382, p. 771.

85. *KSJ* (1959), "Showa 34-nen kaiko," p. 102; *KSJ* (1960), no. 389, p. 71.

86. *KSJ* (1960), no. 394, pp. 238–240; *KSJ* (1960), no. 395, p. 289; *KSJ* (1960), no. 395, pp. 288–291; *KSJ* (1960), no. 396, pp. 333–335; *KSJ* (1960), no. 397, pp. 347–348; *KSJ* (1960), no. 401, pp. 464–466; "Shin Tōkaidō sen Kyoto kōsu rokugatsu ni zen rosen kakutei"; "Kokutetsu Shinkansen hantai dōmei o kessei," *Kyoto shinbun*, March 13, 1960, Yamashina ed., 10.

87. "Shin Tōkaidō sen Kyoto kōsu rokugatsu ni zen rosen kakutei."

88. *KSJ* (1960), no. 395, pp. 287–289.

89. *KSJ* (1960), no. 395, p. 289; *KSJ* (1960), no. 396, p. 333.

90. *KSJ* (1960), no. 396, p. 335.

91. "Kokutetsu Shinkansen hantai dōmei o kessei."

92. *KSJ* (1960), no. 396, p. 334.

93. Nishiguchi, *Shinkansen* (1966), 29.

94. Nishiguchi, *Shinkansen* (1966), 50–51.

95. "Shin Tōkaidō sen 'tōsenbo,'" *Yomiuri shinbun*, morning ed., May 15, 1962, 11.

96. Shimada Kazuo, "Ima wa waraeru Shinkansen umi no kurushimi," *Tabi*, no. 11 (1964): 69, 70.

97. Besides Nishiguchi's *Shinkansen*, see also Shimada Kazuo, "Tanbo no naka no eki," in *Dangan ressha: NHK karā eiga*, no. 8 ([Tokyo]: NHK, 1964), script held at the Tsubouchi Memorial Theatre Museum Library, Waseda University.

98. *KSJ* (1960), no. 401, pp. 466–467.

99. *KSJ* (1961), "Shōwa 36-nen o kaiko shite," pp. 131–132, 138; *KSJ* (1962), "Shōwa 37-nen o kaiko shite," pp. 134–136.

100. They are similar in this way to groups protesting the Anpo revision: Sasaki-Uemura, *Organizing the Spontaneous*, 6.

101. AbdouMaliq Simone, "People as Infrastructure: Intersecting Fragments in Johannesburg," *Public Culture* 16, no. 3 (2004): 407–408.

102. Krauss, "Opposition in Power," 399–400.

103. *KSJ* (1960), no. 389, p. 71.

104. *KSJ* (1960), no. 394, pp. 238–239; *KSJ* (1960), no. 395, pp. 289–290; "Kokutetsu Shinkansen hantai dōmei o kessei"; *KSJ* (1961), "Shōwa 36-nen o kaiko shite," p. 139.

105. "Shi, taisaku kyō tsukuri shori e," *Yomiuri shinbun*, November 18, 1961, Kyoto ed., 12.

106. "Jimotogawa no yōbō hiraku," *Kyoto shinbun*, November 18, 1961, 2.

107. "Jimotogawa no yōbō hiraku"; "Shi, taisaku kyō tsukuri shori e."

108. The English translation of the Constitution of Japan is on the National Diet Library website: https://www.ndl.go.jp/constitution/e/etc/c01.html.

109. "Shi to Kokutetsu ga kyōryoku," *Kyoto shinbun*, city ed., August 8, 1964, 12.

110. Simon Avenell, *Making Japanese Citizens: Civil Society and the Mythology of the* Shimin *in Postwar Japan* (Berkeley: University of California Press, 2010), 83.

111. Avenell, *Making Japanese Citizens*, 148–194.

112. James Scott, *Weapons of the Weak: Everyday Forms of Peasant Resistance* (New Haven, CT: Yale University Press, 1987), xvii.

113. Nakabō Kōhei, *Nakabō Kōhei: Watashi no jikenbo* (Tokyo: Shūeisha, 2000), 20–23.

114. Nakabō, *Nakabō*, 23; "Moto Kokutetsu kakarichō ga kubi tsuri," *Kyoto shinbun*, June 21, 1961, 11; "Kokutetsu oshoku de taiho," *Asahi shinbun*, July 1, 1961, evening ed., 7.

115. Nakabō, *Nakabō*, 24.

116. Nakabō, *Nakabō*, 27–29.

117. Scott, *Weapons of the Weak*, xvi.

118. Nishiguchi, *Shinkansen* (1966), 1.

119. Immanuel Kant, *Metaphysics of Morals*, ed. Lara Denis, trans. Mary Gregor (Cambridge, UK: Cambridge University Press, 2017), 6:437, 203.

120. Nishiguchi, *Shinkansen* (1966), 1.

121. Nishiguchi Katsumi, *Q-to monogatari*, Nishiguchi Katsumi shōsetsushū, vol. 12 (Tokyo: Shin Nihon Shuppansha, 1988), 305.

122. Nishiguchi, *Shinkansen* (1966), 4.

123. Nishiguchi, *Shinkansen* (1966), 4.

124. Nishiguchi, *Shinkansen* (1966), 6–11.

125. Nishiguchi, *Shinkansen* (1966), 13.

126. Nishiguchi, *Shinkansen* (1966), 12. Daikon Genji's surname is related to his occupation (a daikon is a variety of radish), but his given name, written with characters meaning "the next Minamoto," emphasizes his connection to the shrine of the Minamoto progenitor.

127. Nishiguchi, *Shinkansen* (1966), 15.

128. Michael P. Cronin, *Osaka Modern: The City in the Japanese Imaginary* (Cambridge, MA: Harvard University Asia Center, 2017), 14, 9.

129. Nishiguchi, *Shinkansen* (1966), 25.

130. Nishiguchi, *Shinkansen* (1966), 26.

131. Nishiguchi, *Shinkansen* (1966), passim.

132. Nishiguchi, *Shinkansen* (1966), 60.

133. Nishiguchi Katsumi, *Shinkansen*, Nishiguchi Katsumi shōsetsushū, Vol. 6 (Tokyo: Shin Nihon Shuppansha, 1988), 491.

134. Nishiguchi, *Shinkansen* (1988), 491. The book he refers to is discussed in Chapter 3: Tanaka Kakuei, *Nihon rettō kaizō ron* (Tokyo: Nikkan Kogyo Shinbunsha, 1972).

135. Krauss, "Opposition in Power," 411–420.

136. Krauss, "Opposition in Power," 400–414; "Kamogawa o utsukushiku suru kai," https://www.kyoto-kamogawa.jp; Christoph Brumann, *Tradition, Democracy and the Townscape of Kyoto: Claiming a Right to the Past* (New York: Routledge, 2012).

CHAPTER 2

1. Henri Lefebvre, *The Production of Space*, trans. Donald Nicholson-Smith (Cambridge, MA: Blackwell, 1991). For a more theory-focused analysis of the bullet train's impact on space, see Jessamyn R. Abel, "The Power of a Line: How the Bullet Train Transformed Urban Space," *Positions: Asia Critique* 27, no. 3 (2019): 531–555. For a Lefebvrian analysis of 1920s–1930s Tokyo commuter rail, see James A. Fujii, "Intimate Alienation: Japanese Urban Rail and the Commodification of Urban Subjects," *Difference: A Journal of Feminist Cultural Studies* 11, no. 2 (1999): 106–133.

2. Lefebvre, *Production of Space,* 38–42.

3. On the agency of things, see Bruno Latour, *Reassembling the Social: An Introduction to Actor-Network-Theory* (Oxford, UK: Oxford University Press, 2005), 68–72.

4. Brian Larkin, "The Politics and Poetics of Infrastructure," *Annual Review of Anthropology* 42 (August 2013): 327–343.

5. On infrastructures as emergent systems, see Casper Bruun Jensen and Atsuro Morita, "Infrastructures as Ontological Experiments," *Engaging Science, Technology, and Society* 1 (2015): 82–84.

6. Alisa Freedman, "Traversing Tokyo by Subway," in Kären Wigen, Sugimoto Fumiko, and Cary Karacas, eds., *Cartographic Japan*, (Chicago: University of Chicago Press, 2016), 215.

7. "10-nen ato no Tokyo 2: Zadankai (ge)," *Yomiuri shinbun*, January 5, 1959, 8.

8. "10-nen ato no Tokyo 1: Zadankai (jō)," *Yomiuri shinbun*, January 1, 1959, 10. On Takayama, see Carola Hein, "Rebuilding Japanese Cities After 1945," in Carola Hein, Jeffry M. Diefendorf, and Yorifusa Ishida, eds., *Rebuilding Urban Japan After 1945* (New York: Palgrave Macmillan, 2003), 11.

9. Rafael Ivan Pazos Perez, "The Historical Development of the Tokyo Skyline: Timeline and Morphology," *Journal of Asian Architecture and Building Engineering* 13, no. 3 (September 2014): 611.

10. Analyses of this plan in terms of space include Ignacio Adriasola, "Megalopolis and Wasteland: Peripheral Geographies of Tokyo (1961/1971)," *Positions: Asia Critique* 23, no. 2 (May 2015): 206–214; and Jeffrey E. Hanes, "From Megalopolis to Megaroporisu," *Journal of Urban History* 19, no. 2 (February 1993): 66.

11. Perez, "Historical Development," 610–611.

12. Kenzo Tange et al., "A Plan for Tokyo, 1960: Toward a Structural Reorganization," *Japan Architect* 36, no. 4 (April 1961): 17–24. Quotation is from p. 24.

13. "News and Comment," *Japan Architect* 36, no. 4 (1961): 7.

14. Tange et al., "Plan for Tokyo, 1960," 13.

15. Kenzo Tange, "Tokaido-Megalopolis: The Japanese Archipelago in the Future," in Kenzo Tange and Udo Kultermann, eds., *Kenzo Tange, 1946–1969: Architecture and Urban Design* (New York: Praeger, 1970), 165–167. Also see Tange Kenzō, "Nihon rettō no shōraizō: Tōkaidō megaroporisu no keisei," *Chiiki kaihatsu*, no. 2 (November 1964): 8–9. For a broader analysis of the megalopolis idea in Japan, see Hanes, "From Megalopolis to Megaroporisu," 56–94.

16. "Kyōi no hatten 'Tokyo no kōtsūmō,'" *Kyoto shinbun*, September 23, 1964, morning ed., 19; "Shinkansen are kore," *Dōboku gakkai shi*, 49, no. 10 (1964): 42.

17. "Shinkansen are kore," 42.

18. Suzanne Mooney, "Constructed Underground: Exploring Non-Visible Space Beneath the Elevated Tracks of the Yamanote Line," *Japan Forum* 30, no. 2 (2018): 221.

19. Jeffrey Hou, "Vertical Urbanism, Horizontal Urbanity: Notes from East Asian Cities," in Vinayak Bharne, ed., *The Emergent Asian City: Concomitant Urbanities and Urbanisms* (New York: Routledge, 2013), 241.

20. "Shinkansen ato ichi-nen: Zensen 515km o sora kara haiken suru," *Bungei shunjū* 41, no. 11 (1963): [n.p.].

21. "10-nen ato no Tokyo 1: Zadankai (jō)," 10. Also see "Shinkansen are kore," 48.

22. Isomura Eiichi, *Nihon no megaroporisu: Sono jittai to miraizō* (Tokyo: Nihon Keizai Shinbunsha, 1969), 15; Hanes, "From Megalopolis to Megaroporisu," 56. Isomura's work is further examined in Chapter 3.

23. Hanes states that it may have been designed for this purpose (Hanes, "From Megalopolis to Megaroporisu," 74).

24. "Jinkō no suii," *Tokyo-to no tōkei*, https://www.toukei.metro.tokyo.lg.jp /index.htm.

25. "Jūmin kihon daichō jinkō idō hōkoku/Nenpō (Shōsai shūkei)," *e-Stat: Tōkei de miru Nihon*, https://www.e-stat.go.jp.

26. Amano Kōzō, "Tōkaidō Shinkansen no chiiki keizaika ni tsuite," *Un'yu to keizai* 29, no. 2 (1969): 11–17.

27. Chikaraishi Sadakazu, "Improving the Remodelling Plan," trans. Wayne R. Root, *Japan Interpreter* 8, no. 3 (1974): 306. Originally published as "Kaizōron no daian wa kore da," *Chūō kōron* (November 1972), pp. 106–117.

28. Hoshino Yoshirō, "Meishin kōsoku dōro • Shinkansen • Bankokuhaku: Bankokuhaku to Kansai keizai en," *Bessatsu Chūo kōron* 4, no. 4 (December 1965): 198–199.

29. Hoshino, "Meishin kōsoku dōro," 203.

30. For instance, see the contrasting views of Tanaka Kakuei and Isomura Ei-ichi examined in Chapter 3.

31. Both Tange and Hoshino take the latter perspective (Tange, "Nihon rettō no shōraizō," 2–9; and Hoshino, "Meishin kōsoku dōro," 197–206).

32. "Shinkansen are kore," 43.

33. "Shinkansen are kore," 43.

34. "Kiiroi herumetto: Kare nohara ni Shin Osaka Eki," *Osaka jin* 16, no. 3 (1962): 62.

35. The city's design requirements for the new station emphasized the importance of connections to existing and future lines into and through the city (Kokutetsu Osaka Kensen Kōjikyoku, "Sakuhin sakufū: Shin Osaka Eki," *Kenchiku to shakai* 45, no. 10 (October 1964): 18–21). On the location choice and details of urban planning, see Kobayashi Hiroshi, "Shin Osaka Eki shūhen no henbō: Kawariyuku keikan 12," *Chiri* 29, no. 7 (July 1984): 77.

36. Jeffrey E. Hanes, *The City as Subject: Seki Hajime and the Reinvention of Modern Osaka* (Berkeley: University of California Press, 2002), 211.

37. "Chikatetsu ichi-gō-sen Umeda–Shin Osaka eki kaichō ni Shin Yodogawa o wataru," *Osaka shinbun*, September 3, 1964, 7; "Osaka no shin genkan-guchi hiraku," *Osaka shinbun*, September 24, 1964, 7.

38. "Michi ga nai Shin Osaka eki," *Osaka shinbun*, September 8, 1964, 7.

39. "Michi ga nai Shin Osaka eki."

40. "Bōryokudan, haya nawabari arasoi," *Osaka shinbun*, September 14, 1964, 7.

41. Kobayashi, "Shin Osaka Eki shūhen," 80.

42. Kyoto Shikai Jimukyoku Chōsaka, ed., *Kyoto shikai junpō* (hereafter *KSJ*) (1960), no. 392, pp. 161–162; "Shin Tōkaidō-sen ni kechi," *Yomiuri shinbun*, November 27, 1959, evening ed., 7.

43. *KSJ* (1960), no. 392, p. 162.

44. Christopher P. Hood, *Shinkansen: From Bullet Train to Symbol of Modern Japan* (New York: Routledge, 2006), 72–74.

45. "Ichinomiya mo mōhantai," *Asahi shinbun*, January 12, 1961, morning ed., 10. Bisai, an independent city at the time, later merged into Ichinomiya.

46. "Gifu-ken hantai no mama," *Asahi shinbun*, January 12, 1961, morning ed., 11; "Jimoto ga hantai ketsugi," *Yomiuri shinbun*, January 12, 1961, evening ed., 7.

47. "Shin Tōkaidō sen no yōchi baishū," *Yomiuri shinbun*, January 5, 1963, morning ed., 10.

48. Shimada Kazuo, "Ima wa waraeru Shinkansen umi no kurushimi," *Tabi,* no. 11 (1964): 68.   82

49. *KSJ* (1960), no. 391, pp. 138–139; Shimada, "Ima wa waraeru Shinkansen umi no kurushimi," 70; and "Shinkansen are kore," 48.

50. Quoted in Shimada, "Ima wa waraeru Shinkansen umi no kurushimi," 70.

51. Steven J. Ericson, *The Sound of the Whistle: Railroads and the State in Meiji Japan* (Cambridge, MA: Harvard University Asia Center, 1996), 56–57.

52. Carol Gluck calls the railroad a "fact of daily life" by late Meiji, but Ericson qualifies this claim, noting the network's limited reach by 1912 (Carol Gluck, *Japan's Modern Myths: Ideology in the Late Meiji Period* (Princeton, NJ: Princeton University Press, 1985), 101; Ericson, *Sound of the Whistle*, 402, n. 97).

53. Shimada Kazuo, "Tanbo no naka no eki," in *Dangan ressha: NHK karā eiga*, no. 8 ([Tokyo]: NHK, 1964), b9–b10; script held at the Tsubouchi Memorial Theatre Museum Library, Waseda University.

54. Shimada Kazuo, "Madamu to gōrei," in *Dangan ressha: NHK karā eiga*, no. 5 ([Tokyo]: NHK, 1964); script held at the Tsubouchi Memorial Theatre Museum Library, Waseda University.

55. "Nisho, yon keiji," *Kyoto shinbun*, Dai-ichi Shiga ed., February 18, 1962, 10.

56. Yokomitsu Akira, "Kiriha," in *Dangan ressha: NHK karā eiga*, no. 4 ([Tokyo]: NHK, 1964), a2; script held at the Tsubouchi Memorial Theatre Museum Library, Waseda University.

57. Yokomitsu, "Kiriha."

58. Jayson Makoto Chun, *"A Nation of a Hundred Million Idiots"? A Social History of Japanese Television, 1953–1973* (New York: Routledge, 2007), 72–74.

59. "Terebi dorama ni riyō Kokutetsu Shinkansen kōji," *Yomiuri shinbun*, April 1, 1964, evening ed., 10. Also see "Wakiyaku ni tasukerareru," *Yomiuri shinbun*, August 5, 1964, 11; "Shinkansen no kurō hanashi o egaku," *Asahi shinbun*, August 3, 1964, morning ed., 7; and "Yume no chōtokkyū o tēma ni," *Yomiuri shinbun*, July 16, 1964, 10.

60. Several examples are discussed in Chapter 3.

61. Shimada, "Ima wa waraeru Shinkansen umi no kurushimi," 70.

62. Nagasawa Kikuya, *Shinkansen ryokō memo* (Tokyo: Shinju shoin, 1964), 50.

63. Nagasawa, *Shinkansen ryokō memo*, 76.

64. Tim Ingold, *Lines: A Brief History* (New York: Routledge, 2007), 75–84.

65. Wolfgang Schivelbusch, *The Railway Journey: The Industrialization of Time and Space in the 19th Century* (Berkeley: University of California Press, 1986), 37–38, 52–54.

66. Nagasawa, *Shinkansen ryokō memo*, 50.

67. Nagasawa, *Shinkansen ryokō memo*, 50.

68. Kajiyama Toshiyuki, *Yume no chōtokkyū: Chōhen suiri shōsetsu* (Tokyo: Kōbunsha, 1963), 284.

69. Masumura Yasuzō, dir., *Kuro no chōtokkyū* (Daiei, 1964).   ◁○

70. Lefebvre, *Production of Space*, 26.

71. David Seamon, "Body-Subject, Time-Space Routines, and Place-Ballets," in Anne Buttimer and David Seamon, eds., *The Human Experience of Space and Place* (New York: St. Martin's Press, 1980), 148–165.

72. JTB is a private travel company established in November 1963, succeeding a semipublic agency founded in 1912 as Japan Tourist Bureau and renamed Japan Travel Bureau in September 1945.

73. "Chōtokkyū riyō de konna ryokō puran o! Tokyo kara," *Tabi*, no. 11 (1964), 64.

74. "Chōtokkyū riyō de konna ryokō puran o!," 64.

75. "Chōtokkyū riyō de konna ryokō puran o!," 64.

76. "Kōtsū kōsha shusai Shinkansen ryokō puran shū," *Tabi*, no. 11 (1964), 116.

77. "Shasetsu: Biwako ōhashi kansei no igi," *Kyoto shinbun*, September 28, 1964, morning ed., 2.

78. Schivelbusch, *Railway Journey*, 33–42.

79. Ericson, *Sound of the Whistle*, 68–70.

80. Kate McDonald, *Placing Empire: Travel and the Social Imagination in Imperial Japan* (Oakland: University of California Press, 2017).

81. A series of articles in *Asahi* newspaper's Osaka edition aimed at introducing "the real Osakan" was published as a book on the day the bullet train began operations: *Osaka Jin* (Tokyo: Asahi Shinbunsha, 1964).

82. "10-nen ato no Tokyo 11: Kokutetsu (jō)," *Yomiuri shinbun*, January 14, 1959, 8.

83. Tanabe Seiko, "Shinkansen tōjō de miryoku no Izu," *Tabi*, no. 12 (1964): 95. At the time of writing, Tanabe had recently won the prestigious Akutagawa Prize for *Sentimental Journey* (Kanshō ryokō, Senchimentaru jāni), which was written in the Kansai dialect.

84. The discourse of a Tōkaidō megalopolis peaked late in the decade, declining in the early 1970s (based on searches of various databases, including *Asahi shinbun kikuzo II bijuaru*, *Yomidasu rekishikan*, and *Zasshi kiji sakuin shūsei dētabēsu*).

85. Kajiyama Toshiyuki, "Shinkansen kaitsū ato no Osaka," *Tabi*, no. 5 (1965): 62.

86. Kajiyama, "Shinkansen kaitsū ato no Osaka," 62.

87. Kajiyama, "Shinkansen kaitsū ato no Osaka," 62.

88. Schivelbusch, *Railway Journey*, 41–42.

89. On the literary uses of Kansai dialect and other realistic detail in representations of Osaka, see Michael P. Cronin, *Osaka Modern: The City in the Japanese Imaginary* (Cambridge, MA: Harvard University Asia Center, 2017), 150, 178–181.

90. Kajiyama, "Shinkansen kaitsū ato no Osaka," 63.

91. Kajiyama, "Shinkansen kaitsū ato no Osaka," 64.

92. "10-nen ato no Tokyo 1: Zadankai (jō)," 10.
93. On the railroad in Meiji nation-building, see Ericson, *Sound of the Whistle*, 92–94. Regarding the empire, McDonald identifies two views of cultural homogenization among "colonial boosters," who celebrated the "Japanification" of Korea and Manchuria but also made great efforts to maintain differentiation marked by "local color" (McDonald, *Placing Empire*).
94. Yuasa Noriaki, dir., *Daikaijū Gamera* (Daiei, 1965).

CHAPTER 3

1. Fujii Satoshi, *Shinkansen to nashonarizumu* (Tokyo: Asahi Shinbun Shuppan, 2013), 110–132.
2. Kondō Masataka, *Shinkansen to Nihon no hanseiki* (Tokyo: Kōtsū Shinbunsha, 2010), 105–118. Few histories of the bullet train touch on the information society context of the new line's opening. Suda Hiroshi briefly explains how it became "the transport appropriate for businessmen of the information age," but that is in reference to increasing access to information in stations and train cars in the 1980s (Suda Hiroshi, *Tōkaidō Shinkansen sanjū-nen* (Tokyo: Taishō Shuppan, 1994), 201).
3. Langdon Winner, "Do Artifacts Have Politics?" *Daedalus* 109, no. 1 (Winter 1980): 121–122. Winner cites Lewis Mumford, "Authoritarian and Democratic Technics," *Technology and Culture* 5 (1964): 1–8.
4. Tange Kenzō, "Nihon rettō no shōraizō: Tōkaidō megaroporisu no keisei," *Chiiki kaihatsu*, no. 2 (November 1964): 8. Also see Jeffrey E. Hanes, "From Megalopolis to Megaroporisu," *Journal of Urban History* 19, no. 2 (February 1993): 67–70.
5. Umesao describes his development of and publication about the concept in Umesao Tadao, *Jōhō no bunmeigaku* (Tokyo: Chūō kōronsha, 1988), 7, 28.
6. The original publication was in the January 1963 issue of *Hōsō Asahi*. A lightly revised version was published in *Chūō kōron* in March 1963. That version of the article, "Jōhō sangyō ron," is reprinted in Umesao, *Jōhō no bunmeigaku*, 27–52.
7. Itō Yōichi, "Jōhō shakai gainen no keifu to jōhōka ga shūkyō ni oyobosu eikyō," *Tōyō gakujutsu kenkyū tsūshin* 28, no. 3 (1989): 35–36. In English, see Youichi Ito, "Birth of Joho Shakai and Johoka Concepts in Japan and Their Diffusion Outside Japan," *Keio Communication Review*, no. 13 (1991): 4–5.
8. Hayashi Yūjirō, "Jōhō shakai to atarashii kachi taikei," *Chochiku Jihō* 74 (1967): 20–25; and Hayashi Yūjirō, *Johoka Shakai: Hādona shakai kara sofutona shakai e* (Tokyo: Kodansha, 1969). These works are usually taken to mark the height of the information society boom. On these and similar texts, see, for instance, Ito, "Birth of Joho Shakai," 3–12; and Tessa Morris-Suzuki, *Beyond Computopia: Information, Automation, and Democracy in Japan* (New York: Kegan Paul International, 1988).
9. Ito, "Birth of Joho Shakai," 7.
10. Umesao Tadao, "Jōhō sangyō ron saisetsu," in Umesao Tadao, *Jōhō no bunmeigaku* (Tokyo: Chūō kōronsha, 1988), 120.
11. Umesao, "Jōhō sangyō ron saisetsu," 120.

12. "Kamotsu densha mo hashiraseru," *Asahi shinbun*, July 7, 1960, morning ed., 11; "Tōkaidō Shinkansen no kamotsu ressha wa dannen? Kokutetsu," *Asahi shinbun*, January 29, 1965, morning ed., 2; "3 Hours Between Tokyo and Osaka" ([Tokyo]: Japanese National Railways, 1962), in P0.3 "Japan—1963 (July–Dec.): Foreign Participation," Box 276, New York World's Fair 1964–1965 Corporation records, Manuscripts and Archives Division, New York Public Library.

13. Sogō Shinji, "Gijutsu kakumei to Shinkansen," in *Sogō Shinji: Bessatsu* (Tokyo: Sogō Shinji Den Kankōkai, 1988), 148.

14. Umesao Tadao, "Bunmei kaeru chūsū shinkei: Shinkansen no 10-nen," in *Umesao Tadao chosakushū: Nihon kenkyū* (Tokyo: Chūō kōronsha, 1990), 403. Originally published in *Asahi shinbun*, September 20, 1974.

15. Bruce A. Elleman, Elisabeth Köll, and Y. Tak Matsusaka, "Introduction," in Bruce A. Elleman and Stephen Kotkin, eds., *Manchurian Railways and the Opening of China: An International History* (Armonk, NY: Sharpe, 2010), 7.

16. Umesao Tadao, "Jōhō no kōgengaku," in Umesao Tadao, *Jōhō no bunmeigaku* (Tokyo: Chūō kōronsha, 1988), 233.

17. The development of the telegraph inspired such neural metaphors for the media of communication, as well as a similar optimism about both central control and democratic liberation. See Armand Mattelart, *The Invention of Communication*, trans. Susan Emanuel (Minneapolis: University of Minnesota Press, 1996), 47–74.

18. Umesao, "Jōhō no kōgengaku," 234.

19. Mitsuhide Imashiro, "Changes in Japan's Transport Market and JNR Privatization," *Japan Railway and Transport Review*, no. 13 (September 1997): 50.

20. Umesao Tadao, "Jōhō no kōgengaku," 234–235.

21. Katsuji Akita and Yutaka Hasegawa, "History of COMTRAC: Development of the Innovative Traffic-Control System for Shinkansen," *IEEE Annals of the History of Computing* 38, no. 2 (2016): 11–21.

22. Norbert Wiener, *Cybernetics; or, Control and Communication in the Animal and the Machine*, 2nd ed. (Cambridge, MA: MIT Press, 1961).

23. Norbert Wiener, *The Human Use of Human Beings: Cybernetics and Society*, 2nd ed. (Garden City, NY: Doubleday, 1954), 97.

24. Wiener, *Cybernetics*, 160–161.

25. Ronald R. Kline, *The Cybernetics Moment: Or Why We Call Our Age the Information Age* (Baltimore: Johns Hopkins University Press, 2015), 134–137, 192.

26. Wiener's *Human Use of Human Beings* was quickly translated into Japanese and published as *Ningen kikairon: Saibanetikkusu to shakai*, trans. Ikehara Shikao (Tokyo: Misuzu Shobō, 1954). The enthusiastic *Yomiuri* review of this translation quoted here is "Jihyōteki shohyō: Saibanechikkusu," *Yomiuri shinbun*, March 28, 1954, morning ed., 8. Translations of two other books by Wiener (including his autobiography) came out in 1956, and his seminal *Cybernetics; or, Control and Communication in the Animal and the Machine* (New York: Wiley, 1948), which first introduced the concept of cybernetics to the English-reading public, was published

in Japan as *Saibanetikkusu: Dōbutsu to kikai ni okeru seigyo to tsūshin*, trans. Ike-hara Shikao, Iyanaga Shōkichi, and Muroga Saburō (Tokyo: Iwanami Shoten, 1957).

27. James R. Beniger, *The Control Revolution: Technological and Economic Origins of the Information Society* (Cambridge, MA: Harvard University Press, 1986), 21. On protocols and control, see Alexander R. Galloway, *Protocol: How Control Exists After Decentralization* (Cambridge, MA: MIT Press, 2004).

28. Constantine Nomikos Vaporis, *Breaking Barriers: Travel and the State in Early Modern Japan* (Cambridge, MA: Harvard University Press, 1995), 17–55.

29. Jeffrey E. Hanes, *The City as Subject: Seki Hajime and the Reinvention of Modern Osaka* (Berkeley: University of California Press, 2002), 176–193.

30. Media scholar Yuriko Furuhata examines the impact of the "information society" concept on architectural criticism in the 1960s through the works of Tange and other Metabolists in her essay "Architecture as Atmospheric Media: Tange Lab and Cybernetics," in Marc Steinberg and Alexander Zahlten, eds., *Media Theory in Japan* (Durham, NC: Duke University Press, 2017), 52–79.

31. Kenzo Tange, "Recollections: Architect Kenzo Tange (5)," *Japan Architect* 60, no. 8 (August 1985): 7.

32. Rem Koolhaas et al., *Project Japan: Metabolism Talks* (Cologne, Germany: Taschen, 2011), 284, 442–443.

33. Kenzo Tange et al., "A Plan for Tokyo, 1960: Toward a Structural Reorganization," *Japan Architect* 36, no. 4 (April 1961): 10.

34. Tange Kenzō, "Dai Tōa kensetsu kinen eizō keikaku: Kyōgi sekkei tōsen zuan," *Kenchiku zasshi* 56, no. 693 (1942): 963–965.

35. Statistics Bureau, Ministry of Internal Affairs and Communications, "Historical Statistics of Japan," *Statistics Japan*, Chapter 2 Population and Households, http://www.stat.go.jp/english/data/chouki/index.html.

36. Tange, "Nihon rettō no shōraizō," 6–7. A heavily revised version of this article was published several years later in English (as well as German and French) translation as Kenzo Tange, "Tokaido-Megalopolis: The Japanese Archipelago in the Future," in Kenzo Tange and Udo Kultermann, eds., *Kenzo Tange, 1946–1969: Architecture and Urban Design* (New York: Praeger, 1970), 150–167. On the megalopolis concept, see Hanes, "From Megalopolis to Megaroporisu."

37. Tange presented this to the Japan Center for Area Development Research, a group he had recently co-founded as a structure for cooperation for urban and regional planning among academia, industry, and government (Koolhaas et al., *Project Japan*, 678).

38. Tange, "Nihon rettō no shōraizō," 2.

39. Tange, "Nihon rettō no shōraizō," 6.

40. Tange, "Nihon rettō no shōraizō," 6.

41. Tange, "Tokaido-Megalopolis," 154.

42. Hajime Yatsuka, "The 1960 Tokyo Bay Project of Kenzo Tange," in Arie Graafland, ed., *Cities in Transition* (Rotterdam: 010 Publishers, 2001), 178–191; and Furuhata, "Architecture as Atmospheric Media," 58–59.

43. Zhongjie Lin, *Kenzo Tange and the Metabolist Movement: Urban Utopias of Modern Japan* (New York: Routledge, 2010), 90, 95.

44. Quoted in Furuhata, "Architecture as Atmospheric Media," 68. Hyunjung Cho finds a similar example of the Metabolists' presentation of a system of control in terms of individual liberation in Hyunjung Cho, "Expo '70: The Model City of an Information Society," *Review of Japanese Culture and Society* 23 (December 2011): 66.

45. Tange, "Tokaido-Megalopolis," 157.

46. Ravi Ahuja notes that "organicist conceptualisations of social phenomena have the effect of legitimating contradictory social phenomena by naturalising them." (Ravi Ahuja, *Pathways of Empire: Circulation, "Public Works" and Social Space in Colonial Orissa c. 1780–1914* (New Delhi: Orient Black Swan, 2009), 69.)

47. Tanaka Kakuei, *Nihon rettō kaizō ron* (Tokyo: Nikkan Kogyo Shinbunsha, 1972); Kakuei Tanaka, *Building a New Japan: A Plan for Remodeling the Japanese Archipelago* (Tokyo: Simul Press, 1973).

48. Koolhaas et al., *Project Japan*, 649. For these reasons, it is referred to here as his plan.

49. Morris-Suzuki, *Beyond Computopia*, 8–10.

50. Economic Planning Agency, Government of Japan, *New Comprehensive National Development Plan*, ([Tokyo]: 1969), 3.

51. Economic Planning Agency, *New Comprehensive National Development Plan*, 26.

52. Koolhaas et al., *Project Japan*, 13–14, 678; Furuhata, "Architecture as Atmospheric Media," 62.

53. Koolhaas et al., *Project Japan*, 678.

54. Komuta Tetsuhiko, *Tetsudō to kokka: "Gaden intetsu" no kingendaishi* (Tokyo: Kōdansha, 2012), 99–101.

55. Koolhaas et al., *Project Japan*, 649.

56. Tanaka, *Building a New Japan*, 114–122, 166.

57. Hanes, "From Megalopolis to Megaroporisu," 76–77. On Tanaka's background and history of bringing central government resources in the form of infrastructure construction projects to his Niigata district, see Chalmers Johnson, "Tanaka Kakuei, Structural Corruption, and the Advent of Machine Politics in Japan," *Journal of Japanese Studies* 12, no. 1 (1986): 3–11. On his role in bringing high-speed rail to Niigata, see Christopher P. Hood, *Shinkansen: From Bullet Train to Symbol of Modern Japan* (New York: Routledge, 2006), 75–79.

58. Tanaka, *Building a New Japan*, 116.

59. Tanaka, *Building a New Japan*, 121.

60. Tanaka, *Building a New Japan*, 161–164.

61. Tanaka, *Building a New Japan*, 82–93.

62. Tanaka, *Building a New Japan*, 175.

63. See, for instance, economist Shimada Kōichi's and novelist Kajiyama Toshiyuki's concerns discussed in Chapter 2. ("10-nen ato no Tokyo 1: Zadankai (jō),"

Yomiuri shinbun, January 1, 1959, 10; Kajiyama Toshiyuki, "Shinkansen kaitsū ato no Osaka," *Tabi*, no. 5 (1965): 63–64.)

64. In addition to the MIT faculty members listed earlier, Harvard faculty included, among others, Anthony G. Oettinger, a mathematician who founded Harvard's Program on Information Resources Policy in 1972. On Isomura's experience and later use of the ideas he encountered there, see Hanes, "From Megalopolis to Megaroporisu," 56–58.

65. Isomura Eiichi, "A New Proposal for the Relocation of the Capital," trans. Fumiko Bielefeldt and Carl Bielefeldt, *Japan Interpreter* 8, no. 3 (Autumn 1973): 295; originally published as "Shin Tokyo sentoron," *Gekkan ekonomisuto* (November 1972): 48–55.

66. Isomura, "A New Proposal," 300.

67. Isomura, "A New Proposal," 301.

68. Isomura Eiichi, "Shinkansen to tochi kaihatsu riron," *Chiikigaku kenkyū*, no. 3 (September 1973): 45.

69. Isomura, "Shinkansen to tochi kaihatsu riron," 48.

70. Sanuki Toshio, "Toshika jidai no kōtsū taikei," *Un'yu to keizai* 29, no. 2 (February 1969): 53.

71. Isomura, "Shinkansen to tochi kaihatsu riron," 49.

72. Isomura Eiichi, "Toshika no shinten to sono eikyō: Nishi Nihon kaihatsu no kadai," *Chiiki kaihatsu* 7, no. 58 (July 1969): 33.

73. Morris-Suzuki, *Beyond Computopia*, 22.

74. Shunya Yoshimi, "Information," trans. David Buist *Theory, Culture and Society* 23, no. 2–3 (2006), 275–276.

75. Zhongjie Lin makes a similar critique of Tange's 1960 Plan for Tokyo, explaining that the plan failed in part because it implied a strictly hierarchical social structure that conflicted with the ideals of democracy and equality that he claimed for the plan (Lin, *Kenzo Tange and the Metabolist Movement*, 166).

76. William A. Marotti, *Money, Trains, and Guillotines: Art and Revolution in 1960s Japan* (Durham, NC: Duke University Press, 2013), 3–4.

77. Kajiyama Toshiyuki, *Yume no chōtokkyū: Chōhen suiri shōsetsu* (Tokyo: Kōbunsha, 1963); Masumura Yasuzō, dir., *Kuro no chōtokkyū* (Daiei, 1964); Satō Junya, dir., *Shinkansen daibakuha* (Tōei, 1975). Along similar lines, Nishiguchi Katsumi's 1966 novel *Shinkansen* (Tokyo: Kōbundō, 1966), analyzed in Chapter 1, depicted the restriction and selective distribution of information as a means of political manipulation.

78. Morris-Suzuki, *Beyond Computopia*, 16–17.

79. Hayashi, "Jōhō shakai to atarashii kachi taikei," 21. Translation modified from quotation in Ito, "Birth of Joho Shakai," 7.

130   80. Julia Alekseyeva, "Butterflies, Beetles, and Postwar Japan: Semi-documentary in the 1960s," *Journal of Japanese and Korean Cinema* 9, no. 1 (2017): 14–29. Also see Franz Prichard, *Residual Futures: The Urban Ecologies of Literary and Visual Media of 1960s and 1970s Japan* (New York: Columbia University Press,

2019), 20–47; Prichard shows how documentary filmmaker Tsuchimoto Noriaki used images of infrastructures under construction as the basis for a critique of capitalist state power.

81. Masumura Yasuzō, dir., *Kuro no tesuto kā* (Daiei, 1962).

82. Kajiyama, *Yume no chōtokkyū*, 284.

83. Kajiyama, *Yume no chōtokkyū*, 67.

84. Kajiyama, *Yume no chōtokkyū*, 119.

85. Kajiyama, *Yume no chōtokkyū*, 282–283. Though the detective calls himself a "national citizen" (*kokumin*), rather than "citizen" (*shimin*), in this statement Kajiyama echoes a growing sense of the importance of individual action to democracy. See Simon Avenell, *Making Japanese Citizens: Civil Society and the Mythology of the* Shimin *in Postwar Japan* (Berkeley: University of California Press, 2010).

86. Kajiyama, *Yume no chōtokkyū*, 284.

87. "Shinkansen Yokohama eki,'" *Yomiuri shinbun*, October 10, 1962, morning ed., 11.

88. "Giwaku no Nakachi, Nihon Kaihatsu shachō 'kokugai dasse,'" *Yomiuri shinbun*, January 20, 1963, morning ed., 11. Also see "Yōchi baishū de oshoku ka," *Yomiuri shinbun*, October 9, 1962, evening ed., 9; "'Shin Yokohama Eki' de amai shiru," *Asahi shinbun*, November 21, 1962, evening ed., 7.

89. Satō Junya, dir., *Shinkansen daibakuha* (Tōei, 1975). A dubbed version was released in the United States with the title *Bullet Train*, and the film was novelized in English several years later: Joseph Rance and Arei Kato, *Bullet Train* (New York: Morrow, 1980).

CHAPTER 4

1. On the aesthetic function of infrastructure, see Brian Larkin, "The Politics and Poetics of Infrastructure," *Annual Review of Anthropology* 42 (August 2013): 327–343.

2. Paul H. Noguchi, *Delayed Departures, Overdue Arrivals: Industrial Familialism and the Japanese National Railways* (Honolulu: University of Hawai'i Press, 1990), 49–52.

3. Yoshikuni Igarashi, *Bodies of Memory: Narratives of War in Postwar Japanese Culture, 1945–1970* (Princeton, NJ: Princeton University Press, 2000), 3–5. Igarashi discusses the Asia Express and the wartime bullet train as precursors to the New Tōkaidō Line on pages 146–148.

4. Igarashi, *Bodies of Memory*, 131–143; Wesley Sasaki-Uemura, *Organizing the Spontaneous: Citizen Protest in Postwar Japan* (Honolulu: University of Hawai'i Press, 2001), 55–80.

5. T. Fujitani, Geoffrey M. White, and Lisa Yoneyama, "Introduction," in *Perilous Memories: The Asia-Pacific War(s)*, edited by T. Fujitani, Geoffrey M. White, and Lisa Yoneyama (Durham, NC: Duke University Press, 2001), 7.

6. Lisa Yoneyama, *Cold War Ruins: Transpacific Critique of American Justice and Japanese War Crimes* (Durham, NC: Duke University Press, 2016), 1–36.

7. Mariko Asano Tamanoi, *Memory Maps: The State and Manchuria in Postwar Japan* (Honolulu: University of Hawai'i Press, 2009), 5.

8. Ran Zwigenberg, *Hiroshima: The Origins of Global Memory Culture* (New York: Cambridge University Press, 2014), 6.

9. Igarashi, *Bodies of Memory*, 5.

10. Franziska Seraphim, *War Memory and Social Politics in Japan, 1945–2005* (Cambridge, MA: Harvard University Asia Center, 2006), 22.

11. Igarashi, *Bodies of Memory*, 5.

12. Penny Harvey and Hannah Knox, "Enchantments of Infrastructure," *Mobilities* 7, no. 4 (November 2012): 527.

13. On the development of a "victim consciousness" in postwar Japan, see James Joseph Orr, *The Victim as Hero: Ideologies of Peace and National Identity in Postwar Japan* (Honolulu: University of Hawai'i Press, 2001).

14. Aaron S. Moore, *Constructing East Asia: Technology, Ideology, and Empire in Japan's Wartime Era, 1937–1945* (Stanford, CA: Stanford University Press, 2013), 3. Also see Daqing Yang, *Technology of Empire: Telecommunications and Japanese Expansion in Asia, 1883–1945* (Cambridge, MA: Harvard University Asia Center, 2010).

15. Janis Mimura, *Planning for Empire: Reform Bureaucrats and the Japanese Wartime State* (Ithaca, NY: Cornell University Press, 2011), 4–5, 9.

16. Yoshihisa Tak Matsusaka, *The Making of Japanese Manchuria, 1904–1932* (Cambridge, MA: Harvard University Asia Center, 2001), 147–148, 394; Clarence B. Davis and Kenneth E. Wilburn, Jr., eds., *Railway Imperialism* (New York: Greenwood Press, 1991).

17. Louise Young, *Japan's Total Empire: Manchuria and the Culture of Wartime Imperialism* (Berkeley: University of California Press, 1998), 246; Matsusaka, *Making of Japanese Manchuria*, 147–148.

18. Matsusaka, *Making of Japanese Manchuria*, 394–395.

19. Harada Katsumasa, *Mantetsu* (Tokyo: Iwanami Shoten, 1981), 179; Young, *Japan's Total Empire*, 247.

20. Young, *Japan's Total Empire*, 246.

21. Mizukoshi Hiroshi, "Ajia-gō jōmu no omoide," *Tetsudō pikutoriaru* 14, no. 8 (August 1964): 39.

22. For example, Maeda Sei, "'Tokkyū Ajia' o kataru," *Tetsudō no kenkyū* 20, no. 5 (June 15, 1940): 46–47, in Princeton University Department of Rare Books and Special Collections, Cotsen Library, Cohn 200802 Box M1, item 15; and "ASIA," *Manchuria Daily News*, October 23, 1934, 7.

23. Minami Manshū Tetsudō Kabushiki Gaisha, *Ryūsenkei Tokkyū Ajia* (Dalian, China: Minami Manshū Tetsudō, 1936, reprinted 1981), 5; Y-sei, "Mantetsu no hokori 'Ajia,'" *Kagaku Asahi* 2, no. 10 (1942), 28–29. These innovations garnered U.S. government attention: "Report on Manchurian Railway Shops and Rolling Stock," 3245, CHI-125, October 23, 1943, submitted by Charles Layng, Economic Warfare Section, Department of Justice (Chicago), NARA II: 331/8363, General Subject File 1945–52.

24. For example, "Ryokō wa Manshū e!," *Asahi shinbun*, April 6, 1935, morning ed., 24; "Full Text of the F.B.I. Mission's Report," *Japan Times*, January 15, 1935, 3; and "S.M.R.'s New Super-Express Train: 'Asia,'" *Manshū gurafu* 2, no. 6 (November 1934): [n.p.].

25. *Ryūsenkei Tokkyū Ajia* (Dalian, China: Minami Manshū Tetsudō, 1934, reprinted 1981), 1, 6, 10; "S.M.R.'s New Super-Express Train: 'Asia.'"

26. Y-sei, "Mantetsu no hokori 'Ajia,'" 28.

27. "Sen-Man supīdo-appu kinō kara jisshi," *Asahi shinbun*, November 2, 1934, morning ed., 3; "Sekai ichi kaisoku ressha," *Asahi shinbun*, November 12, 1934, morning ed., 13; "Mantetsu de sekai ichi sokuryoku ressha keikaku," *Yomiuri shinbun*, November 12, 1934, morning ed., 7.

28. See, for example, "Ryokō wa Manshū e!"; "Yakushin no Manshū wa maneku," *Asahi shinbun*, March 22, 1936, morning ed., 7; and "Tokkyū 'Ajia' ni muden," *Asahi shinbun*, March 8, 1940, morning ed., 7.

29. Kate McDonald, *Placing Empire: Travel and the Social Imagination in Imperial Japan* (Oakland: University of California Press, 2017), 36–41, 62–66.

30. *Shōgaku kokugo yomihon*, vol. 10; reprinted in Kaigo Tokiomi, ed., *Nihon kyōkasho taikei kindai hen*, vol. 8 (Tokyo: Kōdansha, 1964), 150–153.

31. Harada, *Mantetsu*, 179.

32. *Shōgaku kokugo yomihon*, 153.

33. *Shōgaku kokugo yomihon*, 150.

34. McDonald, *Placing Empire*, 36–41.

35. "Personal and Topical: Miss Diemer Leaves," *Manchuria Daily News*, March 23, 1935, 2; "French Rly. Expert Finds Asia Express Best in the East," *Manchuria Daily News*, March 19, 1936, 1; "Italian Goodwill Mission Welcomed by Mukdenites; Leaves on Trip to Fushuu," *Manchuria Daily News*, May 3, 1938, 8.

36. "Italian Goodwill Mission," 8. On Paulucci, see Elisabetta Tollardo, *Fascist Italy and the League of Nations, 1922–1935* (London: Palgrave Macmillan, 2016), 78–79.

37. "Pressmen Due In on Asia Express," *Manchuria Daily News*, October 10, 1934, 3.

38. "Japanese Add to Manchukuo Railway Lines," *Chicago Daily Tribune*, December 3, 1934, 8; "Crack Trains Link Manchurian Cities," *New York Times*, December 23, 1934, E8.

39. "Japan-China," *March of Time*, vol. 1, ep. 9, December 13, 1935. Accessed via ProQuest, document ID 1822948578.

40. Description of T. Ralph Morton and Jenny Morton, Papers of T. Ralph Morton, 1924–1972. Centre for the Study of World Christianity, University of Edinburgh. GB 3189 CSCNWW11 on Archives Hub, https://archiveshub.jisc.ac.uk/data/gb3189-cscnww11.

41. Ralph Morton, "Grand Tour in Manchukuo," *The Spectator*, no. 5628 (May 8, 1936): 831.

42. This mode of "seeing" Manchuria was promoted to Japanese audiences and recalled in postwar reminiscences: "Sutanpu no tabi: Mantetsu ensen meisho,"

Asahi shinbun, April 6, 1935, morning ed., 24; *Shōgaku kokugo yomihon*, 151; and Oguma Yoneo, "Shijō annai: Tokkyū 'Ajia,'" *Tetsudō pikutoriaru* 9, no. 2 (February 1959): 21, 23.

43. Morton, "Grand Tour in Manchukuo," 831–832.

44. Morton, "Grand Tour in Manchukuo," 832.

45. Ralph Morton, "Travel in Manchoukuo," *Japan Times & Mail*, July 2, 1936, 8. The newspaper used an alternative spelling for "Manchukuo."

46. Raymond Loewy, *The Locomotive* (New York: Studio, 1937), [n.p.].

47. *Tetsudō no kenkyū* 20, no. 5 (June 15, 1940), [n.p.], in Cotsen Children's Library, Cohn 200802 Box M1, item 15.

48. Maeda, "'Tokkyū Ajia' o kataru," 46–47.

49. This echoing of wartime discourse parallels other roughly contemporaneous accounts of life in Manchukuo by former colonists, in contrast to more frank accounts in the 2000s (Ronald Suleski, "Salvaging Memories: Former Japanese Colonists in Manchuria and the Shimoina Project, 2001–12," in Norman Smith, ed., *Empire and Environment in the Making of Manchuria* (Vancouver: UBC Press, 2017), 198–199).

50. On postwar development and technical aid, see David Arase, *Buying Power: The Political Economy of Japan's Foreign Aid* (Boulder, CO: Lynne Rienner, 1995); and Aaron S. Moore, "From 'Constructing' to 'Developing' Asia: Japanese Engineers and the Formation of the Post-Colonial, Cold War Discourse of Development in Asia," in Hiromi Mizuno, Aaron S. Moore, and John DiMoia, eds., *Engineering Asia: Technology, Colonial Development, and the Cold War Order* (London: Bloomsbury, 2018), 85–112.

51. Oguma Yoneo, "Shijō annai: Tokkyū 'Ajia,'" 21–23.

52. "Tokushū: Kyū Minami Manshū Tetsudō," *Tetsudō pikutoriaru* 14, no. 8 (1964).

53. "Tokkyū 'Ajia' kaisō: Manshū o bakushin shita dangan ressha," in Tokyo 12 Chaneru Hōdōbu, ed., *Shōgen: Watashi no Shōwa-shi* (Tokyo: Bungei Shorin, 1969), 124–135.

54. Mainichi Shinbunsha, ed., *Supīdo 100-nen* (Tokyo: Mainichi Shinbunsha, 1969).

55. Ichihara Yoshizumi, Oguma Yoneo, and Nagata Ryūzaburō, eds., *Omoide no Minami Manshū Tetsudō: Shashinshū* (Tokyo: Seibundō Shinkōsha, 1970); and Ichihara Yoshizumi, Oguma Yoneo, Nagata Ryūzaburō, and An'yōji Osamu, eds., *Minami Manshū Tetsudō: "Ajia" to kyaku, kasha no subete* (Tokyo: Seibundō Shinkōsha, 1971).

56. Oguma, "Shijō annai," 21.

57. On "history in the passive voice," see Carol Gluck, "The Idea of Showa: The Japan of Hirohito," *Daedalus* 119, no. 3 (Summer 1990): 12–13.

58. Oguma, "Shijō annai," 21–22.

59. Mizukoshi, "Ajia-gō jōmu no omoide," 38; Oguri Katsuichi, "Ajia-gō unten no omoide," *Tetsudō pikutoriaru* 14, no. 8 (August 1964): 23; and in the opening commentary in "Tokkyū 'Ajia' kaisō," 126.

60. The references to company spirit are in Mizukoshi, "Ajia-gō jōmu no omoide," 38; "Tokkyū 'Ajia' kaisō," 133; and Mainichi, Supīdo, 114–116. The quotation is from Oguri, "Ajia-gō unten," 25.

61. Kubota Masaji, "Ajia-gō sekkei kaiko," Tetsudō pikutoriaru 14, no. 8 (August 1964): 26.

62. Mizukoshi, "Ajia-gō jōmu no omoide," 38.

63. Mizukoshi, "Ajia-gō jōmu no omoide," 38. A similar, if more muted, expression of this view can be seen in Ichihara et al., Minami Manshū Tetsudō, v.

64. Kubota, "Ajia-gō sekkei kaiko," 26.

65. "Tokkyū 'Ajia' kaisō," 134–135; Mainichi, Supīdo, 115–116; and Kubota, "Ajia-gō sekkei kaiko," 26.

66. "Tokkyū 'Ajia' kaisō," 130–131.

67. Mizukoshi, "Ajia-go jōmu no omoide," 38.

68. "Kōbusha kara," Tetsudō pikutoriaru 14, no. 8 (August 1964): 94.

69. "Tokkyū 'Ajia' kaisō," 126.

70. "Tokkyū 'Ajia' kaisō," 133.

71. Harada, Mantetsu, 180.

72. Maema Takanori, Dangan ressha: Maboroshi no Tokyo hatsu Pekin yuki chōtokkyū (Tokyo: Jitsugyō no Nihonsha, 1994), 118–119.

73. Harada Katsumasa, Tetsudō no kataru Nihon no kindai (Tokyo: Soshiete, 1983), 220.

74. The exhibits of the Shōwa Hall present a similar view (Kerry Smith, "The Shōwa Hall: Memorializing Japan's War at Home," Public Historian 24, no. 4 (Autumn 2002): 54–55).

75. Harada Katsumasa, "The Technical Progress of Railways in Japan in Relation to the Policies of the Japanese Government," in Erich Pauer, ed., Papers on the History of Industry and Technology of Japan, Vol. 2 (Marburg, Germany: Marburger Japan-Reihe, 1995), 185–214; Harada, Tetsudō no kataru Nihon no kindai, 220–221.

76. Tetsudōshō Kansen Chōsaka, "Tokyo-Shimonoseki Shinkansen no zōsetsu," Shūhō, no. 192 (June 19, 1940): 35–39. Shūhō was produced by the Cabinet Information Bureau.

77. Tetsudōshō Kansen Chōsaka, "Tokyo-Shimonoseki," 35–36, 39. Strengthening connections to the continent continued earlier transportation policy for the empire. As Kate McDonald notes, as early as 1912, special express trains from Tokyo to Shimonoseki were timed to meet the ferry to Korea (Kate McDonald, "Asymmetrical Integration: Lessons from a Railway Empire," Technology and Culture 56, no. 1 (January 2015): 129).

78. "Tokyo/Shimonoseki no kansen ni kō-A no dangan ressha," Asahi shinbun, October 28, 1939, morning ed., 11; "Senro no haba ga hiroku naru," Asahi shinbun, November 12, 1939, morning ed., 6; and "Shin Tanna suidō jichinsai," Yomiuri shinbun, September 20, 1941, morning ed., 3.

79. See, for instance, "Tunnel to Chosen May Take 20 Years," Japan Times & Advertiser, June 20, 1941, 3.

80. "Dangan ressha iyoyo kidō ni," *Yomiuri shinbun*, June 2, 1939, morning ed., 7. Also see "Senro no haba," 6.

81. Tachibana Jirō, "Tokyo Shimonoseki aida Shinkansen ni tsuite," *Kōgyō zasshi* 76, no. 952 (April 1940): 169. Tachibana included the possibility of a "train ferry," but his point was that shared gauge and other standards would allow train cars to use tracks throughout the empire.

82. Tetsudōshō Kansen Chōsaka, "Tokyo-Shimonoseki," 39.

83. "Nichi-Man-Shi no gijutsu sōryokusen," *Asahi shinbun*, September 28, 1940, evening ed., 2.

84. "Dangan ressha kaisoku ni kakushin," *Asahi shinbun*, October 12, 1940, morning ed., 7; "Trains with 125 Miles an Hour Speed Visualized for New Line," *Japan Times & Mail*, October 15, 1940, 3; quotation is from "Dangan ressha bakushin no jikken e," *Asahi shinbun*, May 29, 1941, evening ed., 2.

85. For example, "Senro no haba," 6; "Dangan ressha no hanashi (1)," *Asahi shinbun*, June 2, 1941, morning ed., 3; "Kōki dangan ressha (2)," *Asahi shinbun*, June 3, 1941, morning ed., 3.

86. "Dangan ressha: Tokyo shichū wa suidō de," *Asahi shinbun*, January 2, 1943, morning ed., 3.

87. Tachibana Jirō, "Kokunai no yusō kyōka e," *Asahi shinbun*, January 26, 1944, morning ed., 4.

88. "Dangan ressha: Tokyo shichū," 3; Arima Hiroshi, "Yakushin suru tetsudō gijutsu," *Asahi shinbun*, October 15, 1942, morning ed., 4.

89. Arima, "Yakushin suru tetsudō gijutsu," 4.

90. James R. Brandon, *Kabuki's Forgotten War: 1931–1945* (Honolulu: University of Hawai'i Press, 2009), 208.

91. "Shin Tanna suidō jichinsai"; "Shin Tanna suidō kyō Atami de kikōshiki," *Yomiuri shinbun*, March 21, 1942, evening ed., 2. Also see "Dangan ressha bakushin no jikken e," *Asahi shinbun*, May 29, 1941, evening ed., 2.

92. "Sensō ga unda kyōshitsu wa Shin Tanna," *Yomiuri shinbun*, June 20, 1944, morning ed., 3.

93. Quoted in "Sensō ga unda kyōshitsu," 3.

94. "Sensō ga unda kyōshitsu," 3.

95. "Sensō ga unda kyōshitsu," 3.

96. Igarashi, *Bodies of Memory*, 12, 16.

97. Smith, "Shōwa Hall," 60. For more recent examples, see Akiko Hashimoto, *The Long Defeat: Cultural Trauma, Memory, and Identity in Japan* (New York: Oxford University Press, 2015).

98. Smith, "Shōwa Hall," 54–55.

99. Tamanoi, *Memory Maps*, 141.

100. Ishida Reisuke, "Jo," in Nihon Kokuyū Tetsudō, ed., *Tōkaidō Shinkansen kōji shi: Doboku hen* ([Tokyo]: Nihon Kokuyū Tetsudō Tōkaidō Shinkansen shisha, 1965), [n.p.].

101. Nihon Kokuyū Tetsudō, *Tōkaidō Shinkansen kōji shi*, 1–5.

102. Mainichi, *Supīdo*, 94–104; quotation is on page 104.
103. Mainichi, *Supīdo*, 102.
104. Mainichi, *Supīdo*, 102.
105. Smith, "Shōwa Hall," 58.
106. Takagi Takeo, "13-nenme no ashioto," *Yomiuri shinbun*, August 15, 1958, morning ed., 7. The population of Japan in 1958 was approximately ninety million.
107. Takagi, "13-nenme no ashioto," 7.
108. Takagi Takeo, "Hakkotsu to Shinkansen," *Yomiuri shinbun*, August 28, 1958, morning ed., 11.
109. Takagi, "Hakkotsu to Shinkansen," 11.
110. Takagi, "Hakkotsu to Shinkansen," 11.
111. Takagi, "Hakkotsu to Shinkansen," 11.
112. "Shinkansen no kurō hanashi o egaku," *Asahi shinbun*, August 3, 1964, morning ed., 7.
113. "Terebi dorama ni riyō Kokutetsu Shinkansen kōji," *Yomiuri shinbun*, April 1, 1964, evening ed., 10.
114. Shimada Kazuo, "Tonneru no kuchō," *Dangan ressha: NHK karā eiga*, no. 2–3 ([Tokyo]: NHK, 1964), a1–a4, d14–d15; script held at the Tsubouchi Memorial Theatre Museum Library, Waseda University.
115. Shimada, "Tonneru," b10.
116. Shimada, "Tonneru," c1–c2.
117. Shimada, "Tonneru," d14–d15.
118. Andrew Gordon, "Managing the Japanese Household: The New Life Movement in Postwar Japan." *Social Politics* 4, no. 2 (1997): 245–283.
119. Shimada, "Tonneru," a5–a8.
120. Shimada, "Tonneru," b1, b4–b5, b7, b10, b13, b16–b18.
121. Shimada, "Tonneru," a19–a20, b1.
122. Shimada, "Tonneru," b19.
123. Shimada, "Tonneru," b14–b15, c15.
124. Shimada, "Tonneru," d5, d12.
125. Namiko Kunimoto, *The Stakes of Exposure: Anxious Bodies in Postwar Japanese Art* (Minneapolis: University of Minnesota Press, 2017), 101–102, 109.
126. "Tetsudō yomoyamabanashi," *Kyoto shinbun*, September 26, 1964, evening ed., 3.
127. "Yume de naku natta 'chōtokkyū,'" *Asahi shinbun*, October 1, 1964, 10.

CHAPTER 5

1. Memo from Antonio de Grassi, Jr., to Charles Poletti, August 27, 1963; memo from Gates Davison to Robert Moses, September 4, 1963; letters from Robert Moses to Kentaro Ayabe, Hajime Fukuda, Masayoshi Ōhira, and Taisuke Ishida, September 6, 1963, in "Japan—1963 (July–Dec.): Foreign Participation," Box 276, New York World's Fair 1964–1965 Corporation records, Manuscripts and Archives Division, New York Public Library. Hereafter NYWF64/65-NYPL.

2. "$375 Million Authorized for Urban Transit Grants," *CQ Almanac 1964* (Washington, DC, 1965): 556–560, http://library.cqpress.com/cqalmanac/cqal64 -1303340.

3. "United States Delegations to International Conferences: ECAFE Study of To-kaido Railway," *Department of State Bulletin* 48, no. 1228 (January 7, 1963), 660; New Jersey State Highway Department, Division of Railroad Transportation, "The Japanese National Railway System and Its New Tokaido Line: A Report," Trenton, NJ: Bureau of Public Information, New Jersey State Highway Department, 1963; "Japanese Train Is Surveyed by U.S. Transport Experts," *New York Times*, April 4, 1965, 42.

4. John W. Dower, *War Without Mercy: Race and Power in the Pacific War* (New York: Pantheon Books, 1986); Naoko Shibusawa, *America's Geisha Ally: Reimagining the Japanese Enemy* (Cambridge, MA: Harvard University Press, 2006). On efforts to change this image through the sister city program, see Meredith Oda, "Masculinizing Japan and Reorienting San Francisco: The Osaka-San Francisco Sister-City Affiliation During the Early Cold War," *Diplomatic History* 41, no. 3 (2017): 460–488.

5. John Swenson-Wright, *Unequal Allies?: United States Security and Alliance Policy Toward Japan, 1945–1960* (Stanford, CA: Stanford University Press, 2005); Sayuri Shimizu, *Creating People of Plenty: The United States and Japan's Economic Alternatives, 1950–1960* (Kent, OH: Kent State University Press, 2001); Nick Kapur, "Mending the 'Broken Dialogue': U.S.-Japan Alliance Diplomacy in the Aftermath of the 1960 Security Treaty Crisis," *Diplomatic History* 41, no. 3 (January 2017): 491–499.

6. Michael Schaller, *Altered States: The United States and Japan Since the Occupation* (New York: Oxford University Press, 1997), 163–183; Kapur, "Mending the 'Broken Dialogue,'" 500–514; Meghan Warner Mettler, "Gimcracks, Dollar Blouses, and Transistors: American Reactions to Imported Japanese Products, 1945–1964," *Pacific Historical Review* 79, no. 2 (2010): 221–229.

7. Michael Adas, *Dominance by Design: Technological Imperatives and America's Civilizing Mission* (Cambridge, MA: Harvard University Press, 2009).

8. Tadokoro Masayuki, "The Model of an Economic Power: Japanese Diplomacy in the 1960s," in Iokibe Makoto, *The Diplomatic History of Postwar Japan*, trans. Robert D. Eldridge (New York: Routledge, 2011), 81–107.

9. David Arase, *Buying Power: The Political Economy of Japan's Foreign Aid* (Boulder, CO: Lynne Rienner, 1995); Aaron S. Moore, "From 'Constructing' to 'Developing' Asia: Japanese Engineers and the Formation of the Post-Colonial, Cold War Discourse of Development in Asia," in Hiromi Mizuno, Aaron S. Moore, and John DiMoia, eds., *Engineering Asia: Technology, Colonial Development, and the Cold War Order* (London: Bloomsbury, 2018), 85–112.

10. Ashley Carse, "Keyword: Infrastructure: How a Humble French Engineering Term Shaped the Modern World," in Penny Harvey, Casper Bruun Jensen, and Atsuro Morita, eds., *Infrastructures and Social Complexity: A Companion* (New York: Routledge, 2017), 33.

11. For example, see Takasaki Tatsunosuke, "Ajia no han'ei to Nihon no unmei," *Chūō kōron* 73, no. 1 (January 1958): 108–109.

12. On "soft power," see Joseph Nye, *Soft Power: The Means to Success in World Politics* (New York: Public Affairs, 2006).

13. The Japanese government thus sought to take advantage of the power of American consumers in international relations. On consumption and U.S. foreign policy, see Christopher Endy, "Travel and World Power: Americans in Europe, 1890–1917," *Diplomatic History* 22, no. 4 (Fall 1998): 565–594; Kristin Hoganson, "Stuff It: Domestic Consumption and the Americanization of the World Paradigm," *Diplomatic History* 30, no. 4 (September 2006): 571–594; and Andrew C. McKevitt, *Consuming Japan: Popular Culture and the Globalizing of 1980s America* (Chapel Hill: University of North Carolina Press, 2017), 4.

14. See, for instance, Eric Dinmore, "Concrete Results? The TVA and the Appeal of Large Dams in Occupation-Era Japan," *Journal of Japanese Studies* 39, no. 1 (2013): 1–38; and Richard J. Samuels, *"Rich Nation, Strong Army": National Security and the Technological Transformation of Japan* (Ithaca, NY: Cornell University Press, 1994).

15. Odd Arne Westad, "The New International History of the Cold War: Three (Possible) Paradigms," *Diplomatic History* 24, no. 4 (2000): 551–565.

16. Kazuhiko Togo, *Japan's Foreign Policy, 1945–2003: The Quest for a Proactive Policy* (Boston: Brill, 2005), 346–347; Tadokoro, "Model of an Economic Power," 86–89.

17. In 1961, the World Bank provided a development loan to JNR, alongside support for highways and other infrastructure projects in Japan: World Bank, *International Bank for Reconstruction and Development (World Bank) annual report 1961–1962* (Washington, DC: World Bank, 1962), 57, http://documents.worldbank .org/curated/en/898261468764426372/International-Bank-for-Reconstruction-and -Development-World-Bank-annual-report-1961-1962.

18. "Closer Tie Between Railways Abroad and JNR," *Japanese Railway Engineering* 3, no. 1 (March 1962): 36. Examples of early industry reporting on the new line include: "You Ought to Know . . . ," *Railway Age*, September 28, 1959, 56; and columns in the series "Railroading After Hours with Jim Lyne," *Railway Age*, October 5, 1959, 28, and May 15, 1961, 53.

19. "Closer Tie Between Railways," 36.

20. Taizo Miyagi, "Post-War Japan and Asianism," *Asia-Pacific Review* 13, no. 2 (2006): 1–16. On Japanese participation in the Bandung Conference, also see Jessamyn R. Abel, *The International Minimum: Creativity and Contradiction in Japan's Global Engagement, 1933–1964* (Honolulu: University of Hawai'i Press, 2015), 218–245.

21. Atsushi Doi, "JNR's Technical Cooperation to Overseas Countries," *Japanese Railway Engineering* 6, no. 2 (June 1965): 5–6.

22. Doi, "JNR's Technical Cooperation," 6.

23. Doi, "JNR's Technical Cooperation," 6.

24. On Japan's accession to ECAFE, see Oba Mie, "Japan's Entry into ECAFE," in Makoto Iokibe, et al., eds., *Japanese Diplomacy in the 1950s: From Isolation to Integration* (New York: Routledge, 2008), 98–113.

25. "Asu kara Tōkaidō Shinkansen kenkyū shūkan," *Asahi shinbun*, April 10, 1964, morning ed., 2; "'Yume no chōtokkyū' o benkyō," *Asahi shinbun*, April 11, 1963, evening ed., 6.

26. Japanese National Railways, *List of Participants: Study Week on New Tokaido Line Project Under Auspices of ECAFE, April 11–18, 1963* ([Tokyo]: Japanese National Railways, 1963).

27. New Jersey State Highway Department, Division of Railroad Transportation, "The Japanese National Railway System and its New Tokaido Line."

28. "Tetsudō Giken o hōmon," *Asahi shinbun*, April 12, 1963, evening ed., 6; "Hanashi no minato," *Yomiuri shinbun*, April 14, 1963, evening ed., 9; "Suteki! Shinjirarenai: Ekafe daihyōdan 'Yume no chōtokkyū' ni shijō," *Asahi shinbun*, April 15, 1963, evening ed., 7; "Shinkansen kōji genba o shisatsu," *Yomiuri shinbun*, April 16, 1963, morning ed., 2; "Ekafe daihyō no mita Kokutetsu," *Asahi shinbun*, April 23, 1963, evening ed., 6; G. M. Magee, "Japan's Tokaido Line: A Report in Depth," *Railway Age* 154, no. 21 (June 3, 1963): 18–22.

29. Quoted in "Suteki! Shinjirarenai," 7.

30. "Hanashi no minato," 9.

31. "United States Delegations to International Conferences," 660.

32. Laurence K. Walrath and Curtis D. Buford, "Report of the United States Delegation to the United Nations Economic Commission for Asia and the Far East: Study Week of the New Tokaido Line, Tokyo, Japan (April 11 to April 18, 1963 Inclusive)," ([1963]), 2; also see Magee, "Japan's Tokaido Line," 18–22.

33. Walrath and Buford, "Report," 21, 27.

34. Walrath and Buford, "Report," 63.

35. "Rails—Japan and Jersey," *New York Times*, July 30, 1963, 28.

36. "Guest Editorials: The Tokaido Line," *Chicago Tribune*, November 13, 1964, 20.

37. "Guest Editorials: The Tokaido Line," 20.

38. Lyndon B. Johnson, "Remarks at the Signing of the High-Speed Ground Transportation Act," September 30, 1965, online by Gerhard Peters and John T. Woolley, *The American Presidency Project*, http://www.presidency.ucsb.edu/ws/?pid=27281; Jim Lyne, "Railroading After Hours," *Railway Age* 154, no. 2 (January 21, 1963): 25.

39. "Japanese Train Is Surveyed by U.S. Transport Experts," 42.

40. "Budd Creates Basic Super-Train Design," *Railway Age* 157, no. 7 (August 24, 1964): 27; "The Pell Plan: Super Railroads for Super Cities," *Railway Age* 157, no. 21 (December 7, 1964): 32–37; "Super-Train of the Future?" *Railway Age* 158, no. 16 (April 19, 1965): 11; Peter Lyon, "Is This Any Way to Ruin a Railroad?" *American Heritage* 19, no. 2 (February 1968), https://www.americanheritage.com/any-way-ruin-railroad.

41. "'Sangyō kankō' no puran," *Asahi shinbun*, December 11, 1957, Tokyo ed., 10; "'Sangyō kankō' ni chikarakobu," *Asahi shinbun*, January 8, 1960, Tokyo ed., 10.

42. Robert Trumbull, "Japan Mixes Work and Play," *New York Times*, February 28, 1965, XX45; Japan Travel Bureau, *Your Technical Tour in Japan* (Tokyo: Nihon Keizai Shinbun, [n.d.]), 6.

43. Trumbull, "Japan Mixes Work and Play," XX45.

44. Japan Travel Bureau, *Your Technical Tour in Japan*, 2.

45. Quoted in Trumbull, "Japan Mixes Work and Play," XX45; and in "JTB to Start Industrial Tour Program in August," *Japan Times*, June 23, 1964, 8.

46. For example, "Now, P&O Offers Two Scenic Routes to the Orient," *Chicago Tribune*, October 27, 1968, G13. Also see advertisements in the *New York Times* on November 3, 1968, 34, and September 20, 1970, 509; and in *Illustrated London News*, February 23, 1974, [n.p.].

47. See Abel, *International Minimum*, 149–151.

48. Tsuneji Hibino, ed., *Meet Japan: A Modern Nation with a Memory* (Rutland, VT: Tuttle, 1966), 46.

49. Hibino, *Meet Japan*, 46.

50. Edward Said, *Orientalism* (New York: Pantheon Books, 1978); David Morley and Kevin Robbins, *Spaces of Identity: Global Media, Electronic Landscapes and Cultural Boundaries* (New York: Routledge, 1995); David S. Roh, Betsy Huang, and Greta A. Niu, *Techno-Orientalism: Imagining Asia in Speculative Fiction, History, and Media* (New Brunswick, NJ: Rutgers University Press, 2015); Toshiya Ueno, "Japanimation and Techno-Orientalism," in Bruce Grenville, ed., *The Uncanny: Experiments in Cyborg Culture* (Vancouver: Arsenal Pulp Press, 2002), 223–236.

51. Lawrence Galton, "Commuting at 1,000 M.P.H.," *New York Times*, October 24, 1965, SM77, 82.

52. Alden P. Armagnac, "Due This Year: Trains That Can Go 150 M.P.H.," *Popular Science* 188, no. 1 (January 1966): 88–92.

53. Joseph Gies, *Wonders of the Modern World* (New York: Crowell, 1966), 15.

54. Gies, *Wonders of the Modern World*, 24.

55. Gies, *Wonders of the Modern World*, 25.

56. Gies, *Wonders of the Modern World*, 25.

57. E. J. Kahn, Jr., "Our Far-Flung Correspondents: The Super-Express of Dreams," *New Yorker* 40, no. 2 (November 21, 1964): 156–162.

58. World's fairs are a rich subject for examining the uses of technology displays. See, for example, Robert H. Haddow, *Pavilions of Plenty: Exhibiting America Abroad in the 1950s* (Washington, DC: Smithsonian Institution Press, 1997); Neil Harris, "All the World a Melting Pot? Japan at American Fairs, 1876–1904," in Akira Iriye, ed., *Mutual Images: Essays in American-Japanese Relations* (Cambridge, MA: Harvard University Press, 1975), 24–54; Robert H. Kargon, Karen Fiss, Morris Low, and Arthur P. Molella, *World's Fairs on the Eve of War: Science, Technology, and Modernity, 1937–1942* (Pittsburgh, PA: University of Pittsburgh Press,

2015); and Robert W. Rydell, John E. Findling, and Kimberly Pelle, *Fair America: World's Fairs in the United States* (Washington, DC: Smithsonian Books, 2000).

59. Nihon Bōeki Shinkōkai, ed., *Nyū Yōku sekai hakurankai sanka hōkokusho 1964–1965* (Tokyo: Nihon Bōeki Shinkōkai, 1966), 3. Also see quotation from JETRO director Takashi Maruo in "50 Lands to Share Fair's Limelight for 50 Reasons," *New York Times*, March 29, 1964, F13. For examples from the press, see "Nyū Yōku Sekaihaku 'shokudō' o 'geisha hausu,'" *Asahi shinbun*, February 22, 1964, evening ed., 6; "Kiryū, dokusha no ran," *Yomiuri shinbun*, May 1, 1964, 3; "Sekaihaku de moteta 'Nihon Shō,'" *Yomiuri shinbun*, October 3, 1964, evening ed., 10.

60. Nihon Bōeki Shinkōkai, *Nyū Yōku sekai hakurankai*, 82.

61. Nihon Bōeki Shinkōkai, *Nyū Yōku sekai hakurankai*, 3.

62. On "self-Orientalism" of the time, see Yuko Kikuchi, *Japanese Modernisation and Mingei Theory: Cultural Nationalism and Oriental Orientalism* (New York: RoutledgeCurzon, 2004); and Aoki Tamotsu, *"Nihon bunka ron" no hen'yō: Sengo Nihon no bunka to aidentitī* (Tokyo: Chūō kōronsha, 1999), 86–117.

63. Christina Klein, *Cold War Orientalism: Asia in the Middlebrow Imagination, 1945–1961* (Berkeley: University of California Press, 2003), 7.

64. Ellen P. Conant, "Japan 'Abroad' at the Chicago Exposition, 1893," in Ellen P. Conant, ed., *Challenging Past and Present: The Metamorphosis of Nineteenth-Century Japanese Art* (Honolulu: University of Hawai'i Press, 2006), 254–280; Ellen P. Conant, "Refractions of the Rising Sun: Japan's Participation in International Exhibitions 1862–1910," in Tomoko Sato and Toshio Watanabe, eds., *Japan and Britain: An Aesthetic Dialogue 1850–1930* (London: Lund Humphries, 1991), 79–92; Angus Lockyer, "Japan at the Exhibition, 1867–1877: From Representation to Practice," in Tadao Umesao, Angus Lockyer, and Kenji Yoshida, eds., *Japanese Civilization in the Modern World XVII: Collection and Representation* (Osaka: National Museum of Ethnology, 2001), 67–76; Yamamoto Sae, *Senjika no banpaku to "Nihon" no hyōshō* (Tokyo: Shinwasha, 2012).

65. Kargon et al., *World's Fairs on the Eve of War*, 70; Akiko Takenaka, "Architecture as Wartime Cultural Diplomacy: The Japanese Pavilion at Paris 1937," in Rika Devos, Alexander Ortenberg, and Vladimir Paperny, eds., *Architecture of Great Expositions 1937–1959: Messages of Peace, Images of War* (Burlington, VT: Ashgate, 2015), 71–80; Yamamoto Sae, Aoki Fujio, Jessica Jordan, and Paul W. Ricketts, "From *The Representation of 'Japan'* in Wartime World's Fairs: Modernists and 'Japaneseness,'" *Review of Japanese Culture and Society* 26 (December 2014): 117–127.

66. "Japanese Pavilion/1962," Seattle Public Library, Century 21 Digital Collection, http://cdm16118.contentdm.oclc.org/cdm/landingpage/collection/p15015coll3.

67. Nihon Bōeki Shinkōkai, *Nyū Yōku sekai hakurankai*, 3; "Nyū Yōku Sekaihaku 'shokudō' o 'geisha hausu,'" 6; "Kiryū, dokusha no ran," 3.

68. Alicia Volk, "From Soft Power to Hard Sell: Japan at American Expositions, 1915–1965," in Nancy E. Green and Christopher Reed, eds., *JapanAmerica: Points of Contact, 1876–1970* (Ithaca, NY: Herbert F. Johnson Museum of Art, Cornell

University, 2016), 80; Oleg Benesch and Ran Zwigenberg, *Japan's Castles: Citadels of Modernity in War and Peace* (Cambridge, UK: Cambridge University Press, 2019), 257; William Peterson, "Shinto and Cherry Blossoms: The Remasculinisation of Japan at the 1964–65 New York World's Fair," in Tets Kimura and Jennifer Anne Harris, eds., *Exporting Japanese Aesthetics: Evolution from Tradition to Cool Japan* (Eastbourne, UK: Sussex Academic Press, 2020), 87–89.

69. Robert Trumbull, "Japan's Pavilion Looks to Future," *New York Times*, May 15, 1964, 23.

70. Quoted in "50 Lands to Share Fair's Limelight for 50 Reasons," F13.

71. Nihon Bōeki Shinkōkai, *Nyū Yōku sekai hakurankai*, 3.

72. Kamekura Yūsaku, "'Hōken' kara 'uchū jidai' e," *Yomiuri shinbun*, April 15, 1964, evening ed., 7.

73. Nihon Bōeki Shinkōkai, *Nyū Yōku sekai hakurankai*, 96–99.

74. Weekly report #71 from Antonio de Grassi, Jr., to William Berns, August 2, 1964, in "Japan—Degrassi Reports 1963–64: Foreign Participation," Box 276, NYWF64/65-NYPL.

75. Memo from Antonio de Grassi, Jr., to Charles Poletti, August 27, 1963; memo from Gates Davison to Robert Moses, September 4, 1963; letters from Robert Moses to Kentaro Ayabe, Hajime Fukuda, Masayoshi Ōhira, and Taisuke Ishida, September 6, 1963, in "Japan—1963 (July–Dec.): Foreign Participation," Box 276, NYWF64/65-NYPL.

76. "Participation in the New York World's Fair 1964–1965," pamphlet from the Japanese Exhibitors' Association, October 1963, in "Japan—Brochures, etc.: Foreign Participation," Box 276, NYWF64/65-NYPL. This item is crossed out in pen in the copy preserved in the archive.

77. Letter from S. Nakanishi to Robert Moses, September 19, 1963; memo from Antonio de Grassi to Allen E. Beach, October 10, 1963, in "Japan—1963 (July–Dec.): Foreign Participation," Box 276, NYWF64/65-NYPL.

78. Time-Life Books, *Official Guide New York World's Fair 1964/1965* (New York: Time, 1964), 140. A list of all items exhibited is in Nihon Bōeki Shinkōkai, *Nyū Yōku sekai hakurankai*, 97–103.

79. Kakumoto Ryōhei, "Sekai ni hokoru 'Hikari' 10 no jiman," *Tabi*, no. 11 (1964), 66.

80. Nihon Bōeki Shinkōkai, *Nyū Yōku sekai hakurankai*, 97–103, 122.

81. Trumbull, "Japan's Pavilion Looks to Future," 23.

82. "Nihon no sekaihaku sanka wa seikō: Seifu daihyō kataru," *Yomiuri shinbun*, May 2, 1964, 5; Nihon Bōeki Shinkōkai, *Nyū Yōku sekai hakurankai*, 112.

83. Trumbull, "Japan's Pavilion Looks to Future," 23.

84. Japan Trade Center, "The World's Fair Report—1964 Season: The Summary of Public Opinion Surveys: Japan Pavilion," pp. 4–16 in "Beikoku shōgyō kankei: Hakurankai kankei: Nyū Yōku Sekai Hakurankai kankei," E'331 3-1-2, Diplomatic Archives, Japan.

85. Japan Trade Center, "The World's Fair Report: Japan Pavilion," 15.

86. Satō Tatsurō, "Akikaze tatsu: Kane mōke shugi ni hihan no koe," *Sekai shūhō* 45, no. 40 (October 6, 1964): 63.

87. Japan Trade Center, "The World's Fair Report: Japan Pavilion," 6.

88. Trumbull, "Japan's Pavilion Looks to Future," 23.

89. Nihon Bōeki Shinkōkai, *Nyū Yōku sekai hakurankai*, 159.

90. "Screen: 'Cry for Happy,'" *New York Times*, March 4, 1961, 16. Earlier films featuring geisha include *Geisha Boy* and *The Barbarian and the Geisha*, both from 1958.

91. Jack Cardiff, dir., *My Geisha* (Paramount Pictures, 1962).

92. For instance, a 1960 episode of the travelogue show *John Gunther's High Road* titled "Japan: The People," which aired on ABC on February 6 and July 23, 1960, included a segment on "a geisha girl school" (Harry Harris, "Screening TV," *Philadelphia Inquirer*, February 10, 1960, 21); and "Japan's Changing Face," on the CBS show *Twentieth Century*, aired in March 1960, featuring "geisha girls reciting Shakespeare" (photo caption in *New York Times*, March 6, 1960, X15). Quotation is from an advertisement (which also highlights the trip past "Mt. Fujiyama") for an American Express tour of "the Mysterious East" in the *New York Times*, March 22, 1964, XX22.

93. On the role of gender in shaping perceptions based on technology, see Adas, *Dominance by Design*, 27.

94. Memo from Gates Davison to Charles Poletti, July 15, 1963, in "Japan—1963 (July–Dec.): Foreign Participation," Box 276, NYWF64/65-NYPL.

95. Memo from Antonio de Grassi to Gates Davison, August 21, 1963, in "Japan—1963 (July–Dec.): Foreign Participation," Box 276, NYWF64/65-NYPL.

96. The House of Japan was described this way in an early version of the booklet "World's Fair A to Z," which was later corrected: see memo from Pete McDonnell to Tony de Grassi, February 24, 1964, in NYWF64/65-NYPL, Box 276, "Japan—1964 (Jan.–June): Foreign Participation." The mistake was echoed, for example, in "World's Fair '64: A Preview," *Newsweek*, January 13, 1964, 44; and "Food at the Fair," *Bon Appétit*, April 1964, 5. A March 10, 1964, reprint of the pamphlet titled "The Fair from A-Z" has the corrected wording: Held at Special Collections Research Center, Henry Madden Library, California State University, Fresno; this pamphlet and the *Newsweek* and *Bon Appétit* articles were accessed through *World's Fairs: A Global History of Expositions*, Adam Matthew Digital. For Japanese reactions, see "Nyū Yōku Sekaihaku 'shokudō' o 'geisha hausu,'" 6; and letter from John T. Mochizuki to World's Fair Division of Operations, December 10, 1963, in "Japan—1963 (July–Dec.): Foreign Participation," Box 276, NYWF64/65-NYPL.

97. Weekly report #99 from Antonio de Grassi, Jr., March 6, 1964, in "Japan—Degrassi Reports 1963–64: Foreign Participation," Box 276, NYWF94–65, NYPL; Memo from Charles Poletti to Officers and Staff of International Affairs and Exhibits, March 2, 1964, in "Japan—1964 (Jan.–June): Foreign Participation," Box 276, NYWF64–65, NYPL.

98. Homer Bigart, "Rain Soaks Crowd," *New York Times*, April 23, 1964, 1; see also Philip H. Dougherty, "4,000 in Opening Parade Slog Bravely Through the Rain with Music and Majorettes," *New York Times*, April 23, 1964, 30.

99. Tadokoro cites *L'Express*, August 8, 1963, 4: Tadokoro, "Model of an Economic Power," 87, 106n5.

100. "Kyō no mondai: Bunka maindo," *Asahi shinbun*, December 26, 1963, evening ed., 1.

101. On requests for Japanese art, see memorandum from Charles Poletti to Tony de Grassi, December 30, 1964, in "Japan—Degrassi Reports 1963–64: Foreign Participation," Box 276, NYWF64/65-NYPL; and all correspondence in "Japan—Art Treasure—1965: Foreign Participation," Box 276, NYWF64/65-NYPL.

102. Japan Trade Center, "The World's Fair Report: Japan Pavilion," 13–17.

103. Floor plans and lists of items on display for both seasons are in Nihon Bōeki Shinkōkai, *Nyū Yōku sekai hakurankai*, 96–103.

104. "House of Japan," *The World's Fair Community*, June 10, 2004, http://www.worldsfaircommunity.org/topic/3029-house-of-japan/.

105. Japan Trade Center, "The World's Fair Report, 1964 Season: Summary of Public Opinion Survey," in "Beikoku shōgyō kankei: Hakurankai kankei: Nyū Yōku Sekai Hakurankai kankei," E'331 3-1-2: 21, Diplomatic Archives, Japan; Nihon Bōeki Shinkōkai, *Nyū Yōku sekai hakurankai*, 96–103.

106. Sanuki Toshio, "Toshika jidai no kōtsū taikei," *Un'yu to keizai* 29, no. 2 (February 1969): 52.

107. Action Memorandum from the Assistant Secretary of State for Far Eastern Affairs (Bundy) to the Under Secretary of State (Ball), November 9, 1964, in *Foreign Relations of the United States* (hereafter *FRUS*), 1964–1968, vol. 29, part 2, eds. Karen L. Gatz and Edward C. Keefer (Washington, DC: U.S. Government Printing Office, 2006), doc. 33, p. 42.

108. "Memorandum From James C. Thomson, Jr., of the National Security Council Staff to President Johnson," January 11, 1965, in *FRUS*, vol. 29, part 2, doc. 40, Tab A, p. 64; Memorandum of Conversation, January 12, 1965, in *FRUS*, vol. 29, part 2, doc. 41, p. 70.

109. This trend was well reported in the American media; see, for example, Joseph Ator, "If You Can't Outsell Japanese, Join 'Em," *Chicago Daily Tribune*, April 20, 1960, C5, C7; George Bliss, "Electronics Firms, Union Act to Fight Jap Imports," *Chicago Daily Tribune*, August 17, 1959, A4; Jonathan Spivak, "Electronics Firms' Profits Margins Are Slipping Despite Sales Rise," *Wall Street Journal*, September 15, 1959, 32.

110. Memorandum of Conversation, January 12, 1965, in *FRUS*, vol. 29, part 2, doc. 41, pp. 72–73.

111. Dana Frank, *Buy American: The Untold Story of Economic Nationalism* (Boston: Beacon Press, 1999) 102–128.

112. Dana Adams Schmidt, "Buy American Act Facing New Test," *New York Times*, April 8, 1960, 14; Richard E. Mooney, "Japan Protests U.S. Action on Bid,"

New York Times, April 29, 1960, 4; "Japanese Given Canal Contract," *New York Times*, May 12, 1960, 16; "Japanese Firm Gets Big Panama Canal Pact," *Chicago Daily Tribune*, May 13, 1960, A7; "You Ought to Know . . . ," *Railway Age*, May 16, 1960, 72.

113. Richard E. Mooney, "Federal Agency Rules Low Bid Doesn't Always Win a Contract," *New York Times*, August 22, 1960, 35.

114. "Third Interior Agency Contract in Two Weeks Goes to Japanese Firm," *Wall Street Journal*, November 28, 1962, 32.

115. Mooney, "Federal Agency Rules," 35.

116. Time-Life Books, *Official Guide*, 143. Hitachi ran many ads featuring the bullet train speeding across bridges or out of tunnels: *New York Times*, January 5, 1964, E7; and several issues of *Railway Age* (January 20, 1964, 74; May 4, 1964, 39; and September 21, 1964, 4). A full-age ad in the journal of the U.S. Chamber of Commerce linked the bullet train to turbines as two Hitachi products representing the company's excellence: *Nation's Business* 52, no. 3 (March 1964): 13.

117. "$375 Million Authorized for Urban Transit Grants."

118. Memorandum of Conversation, July 2, 1964, in *Digital National Security Archive*, doc. JU00331, p. 1; Telegram from Embassy in Japan to Department of State, August 20, 1964, in *FRUS*, vol. 29, part 2, doc. 22, p. 31; Memorandum from William P. Bundy to Secretary of State Dean Rusk, January 6, 1965, in *Digital National Security Archive*, doc. JU00425, p. 4.

119. Robert E. Bedingfield, "U.S. Envisions a Steel Turnpike of Rails for Northeast Corridor," *New York Times*, December 20, 1964, F1; Lyndon B. Johnson, "Annual Message to the Congress on the State of the Union," January 4, 1965, online by Gerhard Peters and John T. Woolley, *The American Presidency Project*, http://www.presidency.ucsb.edu/ws/index.php?pid=26907.

120. Memorandum of Conversation, January 12, 1965, in *FRUS*, vol. 29, part 2, doc. 41, p. 72; Memorandum of Conversation, January 13, 1965, in *FRUS*, vol. 29, part 2, doc. 44, pp. 82–83.

121. Memorandum from Secretary of State Rusk to President Johnson, July 25, 1964, in *FRUS*, vol. 29, part 2, doc. 18, p. 24.

122. Lyndon B. Johnson, "President's Toast at a Dinner in Honor of Prime Minister Sato," January 12, 1965, online by Gerhard Peters and John T. Woolley, *The American Presidency Project*, http://www.presidency.ucsb.edu/ws/?pid=26819.

123. Memorandum of Conversation, January 13, 1965, in *FRUS*, vol. 29, part 2, doc. 44, p. 82.

CONCLUSION

1. Christopher P. Hood, *Shinkansen: From Bullet Train to Symbol of Modern Japan* (New York: Routledge, 2006).

2. J. L. Sert, F. Léger, and S. Giedion, "Nine Points on Monumentality," in S. Giedion, ed., *Architecture You and Me: The Diary of a Development* (Cambridge, MA: Harvard University Press, 1958), 48.

3. Anna Neimark, "The Infrastructural Monument: Stalin's Water Works Under Construction and in Representation," *Future Anterior* 9, no. 2 (Winter 2012): 1.

4. Exhibit label in the Railway Museum, Saitama City, Japan.

5. Sakai Junko, "'Shōwa-yuki' no taimu mashīn," in "Shinkansen to Tokyō," special issue, *Tokyo jin*, no. 282 (May 2010): 40.

6. Kazama Sachiko, "Man for Remodeling the Japanese Archipelago," Mujin-tō Productions, http://www.mujin-to.com/en/artwork/「列島改造人間」シリーズ/. The book is Tanaka Kakuei, *Nihon rettō kaizō ron* (Tokyo: Nikkan Kogyo Shinbunsha, 1972), discussed in Chapter 3. The English translations of the print and series titles are Kazama's.

7. Kazama Sachiko, "A Steam Whistle, Mantetsu-man Manifest," Mujin-tō Productions, http://www.mujin-to.com/en/artwork/汽笛一声（満鉄人現る）/.

8. Kazama Sachiko, "Tatakau āto: Manshūkoku to 15-nen sensō, soshite 'Abe Nihonkoku,'" *Shūkan kin'yōbi* 22, no. 16 (April 18, 2014): 26.

9. Kyoto-shi Sōgō Kikakukyoku Rinia/Hokuriku Shinkansen Yūchi Suishinshitsu, "Rinia o, Kyoto e," https://www.city.kyoto.lg.jp/digitalbook/page/0000000076.html.

10. Kadokawa Daisaku, "Gurōkaru intabyū: Kyoto shichō Kadokawa Daisaku-shi," *Nikkei gurōkaru*, no. 245 (June 2, 2014): 27–28.

11. "JR Central Gives Up on Opening New Maglev Train Service in 2027," *Kyodo News*, July 3, 2020, https://english.kyodonews.net/.

12. "Japan's Maglev Pits New Artery Against the Environment," *Nikkei Asia*, June 27, 2020, https://asia.nikkei.com/.

13. "Mantetsu no 'Ajia' ken'in shita 'Pashina' saiken," *Yomiuri shinbun*, February 19, 1982, evening ed., 3.

14. "Kyū Mantetsu no tokkyū 'Ajia,' gaika kakutoku e 40-nen buri ni gō," *Asahi shinbun*, October 17, 1984, evening ed., 10.

15. "Omoide no mei-SL: Pashina fukkatsu," *Yomiuri shinbun*, February 2, 1985, evening ed., 10; "Fukkatsu maboroshi no SL: 'Pashina' shigatsu ni hashiru yūshi," *Asahi shinbun*, February 3, 1985, morning ed., 22.

16. See, for example, "Mantetsu no 'Ajia' ken'in shita 'Pashina' saiken" and "Omoide no na SL: Pashina fukkatsu."

17. Mariko Asano Tamanoi, *Memory Maps: The State and Manchuria in Postwar Japan* (Honolulu: University of Hawai'i Press, 2009), 156–159.

18. Tamanoi, *Memory Maps*, 160. Also see Yukiko Koga, *Inheritance of Loss: China, Japan, and the Political Economy of Redemption After Empire* (Chicago: University of Chicago Press, 2016).

19. "Kyū Mantetsu no shōki kikansha 'Pashina,' mō ichiryō atta," *Asahi shinbun*, August 18, 2001, morning ed., 26; "Mantetsu 'Ajiagō,' 60-nen buri rikisō e," *Asahi shinbun*, July 16, 2002, morning ed., 9; Hirai Yoshikazu, "Mantetsu no chōtokkyū 'Ajia-gō' ippan kōkai," *Asahi shinbun*, May 20, 2019, evening ed., 9.

20. "Mantetsu Ajia-gō kōkai itsu?" *Asahi shinbun*, April 2, 2008, morning ed., 9.

21. David Morley and Kevin Robbins, *Spaces of Identity: Global Media, Electronic Landscapes and Cultural Boundaries* (New York: Routledge, 1995), 147–173; David S. Roh, Betsy Huang, and Greta A. Niu, *Techno-Orientalism: Imagining Asia in Speculative Fiction, History, and Media* (New Brunswick, NJ: Rutgers University Press, 2015); Toshiya Ueno, "Japanimation and Techno-Orientalism," in Bruce Grenville, ed., *The Uncanny: Experiments in Cyborg Culture* (Vancouver: Arsenal Pulp Press, 2002), 223–236.

22. "Infura shisutemu yushutsu senryaku," *Shushō kantei*, May 17, 2013, https://www.kantei.go.jp/jp/singi/keikyou/dai4/kettei.pdf.

23. On the popularity of "weird Japan" stories, see Ryu Spaeth, "How *The New Yorker* Fell into the 'Weird Japan' Trap," *New Republic*, December 17, 2020, https://newrepublic.com/article/160595/new-yorker-japan-rent-family-fabricated.

24. The apology was widely reported, but the quotation is from Alexandra Ma, "Japanese Rail Company Issued an Official Apology After a Train Left 25 Seconds Early," *Business Insider*, May 15, 2018, https://www.businessinsider.com/japan-rail-company-apologises-for-train-leaving-25-seconds-early-2018-5.

Bibliography

ARCHIVES, MUSEUMS, AND DIGITAL COLLECTIONS
American Presidency Project (http://www.presidency.ucsb.edu)
Cotsen Children's Library, Princeton University Department of Rare Books and
    Special Collections, Princeton, NJ
CQ Almanac (http://library.cqpress.com/cqalmanac/)
Digital National Security Archive Collection: Japan and the U.S., 1960–1976
    (Proquest)
Diplomatic Archives of the Ministry of Foreign Affairs of Japan, Tokyo, Japan
Gordon W. Prange Collection, University of Maryland, College Park, MD
Kyoto Prefectural Library and Archives, Kyoto, Japan
National Archives and Records Administration II, College Park, MD
New York World's Fair 1964–1965 Corporation records, Manuscripts and Archives
    Division, New York Public Library, New York, NY
Railway Museum, Saitama City, Japan
Tsubouchi Memorial Theatre Museum Library, Waseda University, Tokyo, Japan
World's Fair Community (http://www.worldsfaircommunity.org)
World's Fairs: A Global History of Expositions, Adam Matthew Digital

NEWSPAPERS AND MAGAZINES
Asahi shinbun
Business Insider
Chicago Tribune
Japanese Railway Engineering
Japan Times
Kyodo News

Kyoto shinbun
Manchuria Daily News
New York Times
Nikkei Asia
Osaka jin
Osaka shinbun
Philadelphia Inquirer
Railway Age
Tabi
Tetsudō pikutoriaru
Wall Street Journal
Yomiuri shinbun

OTHER WORKS CITED

Abel, Jessamyn R. *The International Minimum: Creativity and Contradiction in Japan's Global Engagement, 1933–1964.* Honolulu: University of Hawai'i Press, 2015.

———. "The Power of a Line: How the Bullet Train Transformed Urban Space." *Positions: Asia Critique* 27, no. 3 (2019): 531–555.

Adas, Michael. *Dominance by Design: Technological Imperatives and America's Civilizing Mission.* Cambridge, MA: Harvard University Press, 2009.

———. *Machines as the Measure of Men: Science, Technology, and Ideologies of Western Dominance.* Ithaca, NY: Cornell University Press, 1989.

Adriasola, Ignacio. "Megalopolis and Wasteland: Peripheral Geographies of Tokyo (1961/1971)." *Positions: Asia Critique* 23, no. 2 (May 2015): 201–229.

Aguiar, Marian. *Tracking Modernity: India's Railway and the Culture of Mobility.* Minneapolis: University of Minnesota Press, 2011.

Ahuja, Ravi. *Pathways of Empire: Circulation, "Public Works" and Social Space in Colonial Orissa c. 1780–1914* (New Delhi: Orient Black Swan, 2009).

Akita, Katsuji, and Yutaka Hasegawa. "History of COMTRAC: Development of the Innovative Traffic-Control System for Shinkansen." *IEEE Annals of the History of Computing* 38, no. 2 (2016): 11–21.

Alekseyeva, Julia. "Butterflies, Beetles, and Postwar Japan: Semi-documentary in the 1960s." *Journal of Japanese and Korean Cinema* 9, no. 1 (2017): 14–29.

Amakawa Akira. "The Making of the Postwar Local Government System." In *Democratizing Japan: The Allied Occupation,* edited by Robert E. War and Sakamoto Yoshikazu, 251–282. Honolulu: University of Hawai'i Press, 1987.

Amano Kōzō. "Tōkaidō Shinkansen no chiiki keizaika ni tsuite." *Un'yu to keizai* 29, no. 2 (1969): 11–17.

Anand, Nikhil. *Hydraulic City: Water and the Infrastructures of Citizenship in Mumbai.* Durham, NC: Duke University Press, 2017.

Aoki Tamotsu. *"Nihon bunka ron" no hen'yō: Sengo Nihon no bunka to aidentitī.* Tokyo: Chūō kōronsha, 1999.

Appel, Hannah, Nikhil Anand, and Akhil Gupta. "Introduction: Temporality, Politics, and the Promise of Infrastructure." In *The Promise of Infrastructure*, edited by Nikhil Anand, Akhil Gupta, and Hannah Appel, 1–28. Durham, NC: Duke University Press, 2018.

Arase, David. *Buying Power: The Political Economy of Japan's Foreign Aid*. Boulder, CO: Lynne Rienner, 1995.

Armagnac, Alden P. "Due This Year: Trains That Can Go 150 M.P.H." *Popular Science* 188, no. 1 (January 1966): 88–92.

Avenell, Simon. *Making Japanese Citizens: Civil Society and the Mythology of the Shimin in Postwar Japan*. Berkeley: University of California Press, 2010.

Benesch, Oleg, and Ran Zwigenberg. *Japan's Castles: Citadels of Modernity in War and Peace*. Cambridge, UK: Cambridge University Press, 2019.

Beniger, James R. *The Control Revolution: Technological and Economic Origins of the Information Society*. Cambridge, MA: Harvard University Press, 1986.

Benjamin, Walter. "Theses on the Philosophy of History." In *Illuminations*, edited by Hannah Arendt, translated by Harry Zohn, 253–264. New York: Schocken Books, 1968.

Brandon, James R. *Kabuki's Forgotten War: 1931–1945*. Honolulu: University of Hawai'i Press, 2009.

Brumann, Christoph. *Tradition, Democracy and the Townscape of Kyoto: Claiming a Right to the Past*. New York: Routledge, 2012.

Cardiff, Jack, dir. *My Geisha*. Paramount Pictures, 1962.

Carse, Ashley. "Keyword: Infrastructure: How a Humble French Engineering Term Shaped the Modern World." In *Infrastructures and Social Complexity: A Companion*, edited by Penny Harvey, Casper Bruun Jensen, and Atsuro Morita, 27–39. New York: Routledge, 2017.

Chikaraishi Sadakazu. "Improving the Remodelling Plan." Translated by Wayne R. Root. *Japan Interpreter* 8, no. 3 (1974): 304–322. Originally published as "Kaizōron no daian wa kore da." *Chūō kōron* (November 1972): 106–117.

Cho, Hyunjung. "Expo '70: The Model City of an Information Society." *Review of Japanese Culture and Society* 23 (December 2011): 57–71.

Chun, Jayson Makoto. *"A Nation of a Hundred Million Idiots"? A Social History of Japanese Television, 1953–1973*. New York: Routledge, 2007.

Coleman, Leo. *A Moral Technology: Electrification as Political Ritual in New Delhi*. Ithaca, NY: Cornell University Press, 2017.

Conant, Ellen P. "Japan 'Abroad' at the Chicago Exposition, 1893." In *Challenging Past and Present: The Metamorphosis of Nineteenth-Century Japanese Art*, edited by Ellen P. Conant, 254–280. Honolulu: University of Hawai'i Press, 2006.

———. "Refractions of the Rising Sun: Japan's Participation in International Exhibitions 1862–1910." In *Japan and Britain: An Aesthetic Dialogue 1850–1930*, edited by Tomoko Sato and Toshio Watanabe, 79–92. London: Lund Humphries, 1991.

Constitution of Japan. National Diet Library. https://www.ndl.go.jp/constitution/e/etc/c01.html.

Cronin, Michael P. *Osaka Modern: The City in the Japanese Imaginary.* Cambridge, MA: Harvard University Asia Center, 2017.

Davis, Clarence B., and Kenneth E. Wilburn, Jr., eds. *Railway Imperialism.* New York: Greenwood Press, 1991.

Dinmore, Eric. "Concrete Results? The TVA and the Appeal of Large Dams in Occupation-Era Japan." *Journal of Japanese Studies* 39, no. 1 (2013): 1–38.

———. "'Mountain Dream' or the 'Submergence of Fine Scenery'? Japanese Contestations over the Kurobe Number Four Dam, 1920–1970." *Water History* 6, no. 4 (2014): 315–340.

Dower, John W. *War Without Mercy: Race and Power in the Pacific War.* New York: Pantheon Books, 1986.

Dusinberre, Martin. *Hard Times in the Hometown: A History of Community Survival in Modern Japan.* Honolulu: University of Hawai'i Press, 2012.

Economic Planning Agency, Government of Japan. *New Comprehensive National Development Plan.* [Tokyo]: 1969.

Elleman, Bruce A., Elisabeth Köll, and Y. Tak Matsusaka. "Introduction." In *Manchurian Railways and the Opening of China: An International History*, edited by Bruce A. Elleman and Stephen Kotkin, 3–10. Armonk, NY: Sharpe, 2010.

Endy, Christopher. "Travel and World Power: Americans in Europe, 1890–1917." *Diplomatic History* 22, no. 4 (Fall 1998): 565–594.

Ericson, Steven J. *The Sound of the Whistle: Railroads and the State in Meiji Japan.* Cambridge, MA: Harvard University Asia Center, 1996.

Fisch, Michael. *An Anthropology of the Machine: Tokyo's Commuter Train Network.* Chicago: University of Chicago Press, 2018.

Frank, Dana. *Buy American: The Untold Story of Economic Nationalism.* Boston: Beacon Press, 1999.

Freedman, Alisa. *Tokyo in Transit: Japanese Culture on the Rails and Road.* Stanford, CA: Stanford University Press, 2011.

———. "Traversing Tokyo by Subway." In *Cartographic Japan*, edited by Kären Wigen, Sugimoto Fumiko, and Cary Karacas, 214–217. Chicago: University of Chicago Press, 2016.

Fujii, James A. "Intimate Alienation: Japanese Urban Rail and the Commodification of Urban Subjects." *Difference: A Journal of Feminist Cultural Studies* 11, no. 2 (1999): 106–133.

Fujii Satoshi. *Shinkansen to nashonarizumu.* Tokyo: Asahi Shinbun Shuppan, 2013.

Fujitani, T., Geoffrey M. White, and Lisa Yoneyama. "Introduction." In *Perilous Memories: The Asia-Pacific War(s)*, edited by T. Fujitani, Geoffrey M. White, and Lisa Yoneyama, 1–29. Durham, NC: Duke University Press, 2001.

Furuhata, Yuriko. "Architecture as Atmospheric Media: Tange Lab and Cybernetics." In *Media Theory in Japan*, edited by Marc Steinberg and Alexander Zahlten, 52–79. Durham, NC: Duke University Press, 2017.

Galloway, Alexander R. *Protocol: How Control Exists After Decentralization.* Cambridge, MA: MIT Press, 2004.

Gatz, Karen L., and Edward C. Keefer, eds. *Foreign Relations of the United States, 1964–1968*, vol. 29, part 2. Washington, DC: U.S. Government Printing Office, 2006.

Gies, Joseph. *Wonders of the Modern World*. New York: Crowell, 1966.

Gluck, Carol. "The Idea of Showa: The Japan of Hirohito." *Daedalus* 119, no. 3 (Summer 1990): 1–26.

———. *Japan's Modern Myths: Ideology in the Late Meiji Period*. Princeton, NJ: Princeton University Press, 1985.

Gordon, Andrew. *Fabricating Consumers: The Sewing Machine in Modern Japan*. Berkeley: University of California Press, 2011.

———. "Managing the Japanese Household: The New Life Movement in Postwar Japan." *Social Politics* 4, no. 2 (1997): 245–283.

Haddow, Robert H. *Pavilions of Plenty: Exhibiting America Abroad in the 1950s*. Washington, DC: Smithsonian Institution Press, 1997.

Hamaguchi, Ryuichi. "The Tokaido New Trunk Line." Part 1. *Japan Architect*, no. 110 (July 1965): 12–27.

Hanes, Jeffrey E. *The City as Subject: Seki Hajime and the Reinvention of Modern Osaka*. Berkeley: University of California Press, 2002.

———. "From Megalopolis to Megaroporisu." *Journal of Urban History* 19, no. 2 (February 1993): 56–94.

Harada Katsumasa. *Mantetsu*. Tokyo: Iwanami Shoten, 1981.

———. "The Technical Progress of Railways in Japan in Relation to the Policies of the Japanese Government." In *Papers on the History of Industry and Technology of Japan*, Vol. 2, edited by Erich Pauer, 185–214. Marburg, Germany: Marburger Japan-Reihe, 1995.

———. *Tetsudō no kataru Nihon no kindai*. Tokyo: Soshiete, 1983.

Harris, Neil. "All the World a Melting Pot? Japan at American Fairs, 1876–1904." In *Mutual Images: Essays in American-Japanese Relations*, edited by Akira Iriye, 24–54. Cambridge, MA: Harvard University Press, 1975.

Harvey, Penny, and Hannah Knox. "Enchantments of Infrastructure." *Mobilities* 7, no. 4 (November 2012): 521–536.

Hashimoto, Akiko. *The Long Defeat: Cultural Trauma, Memory, and Identity in Japan*. New York: Oxford University Press, 2015.

Hayashi Yūjirō. *Jōhōka shakai: Hādona shakai kara sofutona shakai e*. Tokyo: Kōdansha, 1969.

———. "Jōhō shakai to atarashii kachi taikei." *Chochiku jihō* 74 (1967): 20–25.

Hein, Carola. "Rebuilding Japanese Cities After 1945." In *Rebuilding Urban Japan After 1945*, edited by Carola Hein, Jeffry M. Diefendorf, and Yorifusa Ishida, 1–16. New York: Palgrave Macmillan, 2003.

Hibino, Tsuneji, ed. *Meet Japan: A Modern Nation with a Memory*. Rutland, VT: Tuttle, 1966.

Hoganson, Kristin. "Stuff It: Domestic Consumption and the Americanization of the World Paradigm." *Diplomatic History* 30, no. 4 (September 2006): 571–594.

Hood, Christopher P. *Shinkansen: From Bullet Train to Symbol of Modern Japan.* New York: Routledge, 2006.

Hoshino Yoshirō. "Meishin kōsoku dōro • Shinkansen • Bankokuhaku: Bankoku-haku to Kansai keizai en." *Bessatsu Chūo kōron* 4, no. 4 (December 1965): 197–206.

Hou, Jeffrey. "Vertical Urbanism, Horizontal Urbanity: Notes from East Asian Cities." In *The Emergent Asian City: Concomitant Urbanities and Urbanisms,* edited by Vinayak Bharne, 234–243. New York: Routledge, 2013.

Ichihara Yoshizumi, Oguma Yoneo, and Nagata Ryūzaburō, eds. *Omoide no Minami Manshū Tetsudō: Shashinshū.* Tokyo: Seibundō Shinkōsha, 1970.

Ichihara Yoshizumi, Oguma Yoneo, Nagata Ryūzaburō, and An'yōji Osamu, eds. *Minami Manshū Tetsudō: "Ajia" to kyaku, kasha no subete.* Tokyo: Seibundō Shinkōsha, 1971.

Igarashi, Yoshikuni. *Bodies of Memory: Narratives of War in Postwar Japanese Culture, 1945–1970.* Princeton, NJ: Princeton University Press, 2000.

Imashiro, Mitsuhide. "Changes in Japan's Transport Market and JNR Privatization." *Japan Railway and Transport Review,* no. 13 (September 1997): 50–53.

"Infura shisutemu yushutsu senryaku." *Shushō kantei,* May 17, 2013. https://www.kantei.go.jp/jp/singi/keikyou/dai4/kettei.pdf.

Ingold, Tim. *Lines: A Brief History.* New York: Routledge, 2007.

Ishida Reisuke. "Jo." In *Tōkaidō Shinkansen kōji shi: Doboku hen,* edited by Nihon Kokuyū Tetsudō, 1–5. [Tokyo]: Nihon Kokuyū Tetsudō Tōkaidō Shinkansen shisha, 1965.

Isomura Eiichi. "A New Proposal for the Relocation of the Capital." Translated by Fumiko Bielefeldt and Carl Bielefeldt. *Japan Interpreter* 8, no. 3 (Autumn 1973): 292–303. Originally published as "Shin Tokyo sentoron," *Gekkan ekonomisuto* (November 1972): 48–55.

———. *Nihon no megaroporisu: Sono jittai to miraizō.* Tokyo: Nihon Keizai Shinbunsha, 1969.

———. "Shinkansen to tochi kaihatsu riron." *Chiikigaku kenkyū,* no. 3 (September 1973): 45–52.

———. "Toshika no shinten to sono eikyō: Nishi Nihon kaihatsu no kadai." *Chiiki kaihatsu* 7, no. 58 (July 1969): 27–38.

Itō Yōichi. "Jōhō shakai gainen no keifu to jōhōka ga shūkyō ni oyobosu eikyō." *Tōyō gakujutsu kenkyū tsūshin* 28, no. 3 (1989): 33–51.

Ito, Youichi. "Birth of Joho Shakai and Johoka Concepts in Japan and Their Diffusion Outside Japan." *Keio Communication Review,* no. 13 (1991): 3–12.

"Japan-China," *March of Time,* vol. 1, ep. 9, December 13, 1935. Accessed via Pro-Quest, document ID 1822948578.

Japanese National Railways. *List of Participants: Study Week on New Tokaido Line Project Under Auspices of ECAFE, April 11–18, 1963.* [Tokyo]: Japanese National Railways, 1963.

"Japanese Pavilion/1962." Seattle Public Library, Century 21 Digital Collection. http://cdm16118.contentdm.oclc.org/cdm/landingpage/collection/p15015coll3.

Japan Travel Bureau. *Your Technical Tour in Japan*. Tokyo: Nihon Keizai Shinbun, [n.d.].

Jensen, Casper Bruun, and Atsuro Morita. "Infrastructures as Ontological Experiments." *Engaging Science, Technology, and Society* 1 (2015): 81–87.

"Jinkō no suii." *Tokyo-to no tōkei*. https://www.toukei.metro.tokyo.lg.jp/index.htm.

Johnson, Chalmers. "Tanaka Kakuei, Structural Corruption, and the Advent of Machine Politics in Japan." *Journal of Japanese Studies* 12, no. 1 (1986): 1–28.

"Jūmin kihon daichō jinkō idō hōkoku/Nenpō (Shōsai shūkei)." *e-Stat: Tōkei de miru Nihon*. https://www.e-stat.go.jp.

Kadokawa Daisaku. "Gurōkaru intabyū: Kyoto shichō Kadokawa Daisaku-shi." *Nikkei Gurōkaru*, no. 245 (June 2, 2014): 26–28.

Kajiyama Toshiyuki. *Yume no chōtokkyū: Chōhen suiri shōsetsu*. Tokyo: Kōbunsha, 1963.

Kamogawa o utsukushiku suru kai. https://www.kyoto-kamogawa.jp.

Kant, Immanuel. *Metaphysics of Morals*, edited by Lara Denis, translated by Mary Gregor. Cambridge, UK: Cambridge University Press, 2017.

Kapur, Nick. "Mending the 'Broken Dialogue': U.S.-Japan Alliance Diplomacy in the Aftermath of the 1960 Security Treaty Crisis." *Diplomatic History* 41, no. 3 (January 2017): 489–517.

Kargon, Robert H., Karen Fiss, Morris Low, and Arthur P. Molella. *World's Fairs on the Eve of War: Science, Technology, and Modernity, 1937–1942*. Pittsburgh, PA: University of Pittsburgh Press, 2015.

Kazama Sachiko. "A Steam Whistle, Mantetsu-man Manifest." Mujin-tō Productions. http://www.mujin-to.com/en/artwork/汽笛一声（満鉄人現る）/.

———. "Man for Remodeling the Japanese Archipelago." Mujin-tō Productions. http://www.mujin-to.com/en/artwork/「列島改造人間」シリーズ/.

———. "Tatakau āto: Manshūkoku to 15-nen sensō, soshite 'Abe Nihonkoku'." *Shūkan kin'yōbi* 22, no. 16 (April 18, 2014): 26–27.

Kahn, E. J., Jr. "Our Far-Flung Correspondents: The Super-Express of Dreams." *New Yorker* 40, no. 2 (November 21, 1964): 155–162.

Kikuchi, Yuko. *Japanese Modernisation and Mingei Theory: Cultural Nationalism and Oriental Orientalism*. New York: RoutledgeCurzon, 2004.

Klein, Christina. *Cold War Orientalism: Asia in the Middlebrow Imagination, 1945–1961*. Berkeley: University of California Press, 2003.

Kline, Ronald R. *The Cybernetics Moment: Or Why We Call Our Age the Information Age*. Baltimore: Johns Hopkins University Press, 2015.

Kobayashi Hiroshi. "Shin Osaka Eki shūhen no henbō: Kawariyuku keikan 12." *Chiri* 29, no. 7 (July 1984): 77–81.

Koga, Yukiko. *Inheritance of Loss: China, Japan, and the Political Economy of Redemption After Empire*. Chicago: University of Chicago Press, 2016.

Kokutetsu Osaka Kensen Kōjikyoku. "Sakuhin sakufū: Shin Osaka Eki." *Kenchiku to shakai* 45, no. 10 (October 1964): 18–21.

Komuta Tetsuhiko. *Tetsudō to kokka: "Gaden intetsu" no kindgendaishi*. Tokyo: Kōdansha, 2012.

Kondō Masataka. *Shinkansen to Nihon no hanseiki.* Tokyo: Kōtsū Shinbunsha, 2010.

Koolhaas, Rem, Hans Ulrich Obrist, Kayoko Ota, James Westcott, and Office for Metropolitan Architecture (AMO). *Project Japan: Metabolism Talks.* Cologne, Germany: Taschen, 2011.

Koyama Seiji, dir. *Tōkaidō Shinkansen.* Shin Riken Eiga, 1964.

Koyama, Tōru. "The Shinkansen (Bullet Train): A New Era in Railway Technology." In *A Social History of Science and Technology in Contemporary Japan,* vol. 3, edited by Shigeru Nakayama and Kunio Goto, 379–389. Melbourne: Trans Pacific Press, 2006.

Krauss, Ellis S. "Opposition in Power: The Development and Maintenance of Leftist Government in Kyoto Prefecture." In *Political Opposition and Local Politics in Japan,* edited by Ellis S. Krauss and Scott E. Flanagan, 383–424. Princeton, NJ: Princeton University Press, 1981.

Kunimoto, Namiko. *The Stakes of Exposure: Anxious Bodies in Postwar Japanese Art.* Minneapolis: University of Minnesota Press, 2017.

Kushner, Barak. *Slurp! A Social and Culinary History of Ramen—Japan's Favorite Noodle Soup.* Boston: Global Oriental, 2012.

"Kyoto Kokusai bunka kankō toshi kensetsu hō" (1950). The text of this and similar laws directed toward other cities is available through *e-Gov: Denshi seifu no sōgō madoguchi.* https://www.e-gov.go.jp.

"Kyoto shimin kenshō." *Kyōtoshi jōhōkan.* https://www.city.kyoto.lg.jp/sogo/page/0000184650.html.

Kyoto Shikai Jimukyoku Chōsaka, ed. *Kyoto shikai junpō.* Kyoto: Kyoto Shikai Jimukyoku Chōsaka, 1958–1964.

Kyoto-shi Sōgō Kikakukyoku Rinia/Hokuriku Shinkansen Yūchi Suishinshitsu. "Rinia o, Kyoto e." https://www.city.kyoto.lg.jp/digitalbook/page/0000000076.html.

Larkin, Brian. "The Politics and Poetics of Infrastructure." *Annual Review of Anthropology* 42 (August 2013): 327–343.

———. "Promising Forms: The Political Aesthetics of Infrastructure." In *The Promise of Infrastructure,* edited by Nikhil Anand, Akhil Gupta, and Hannah Appel, 175–202. Durham, NC: Duke University Press, 2018.

Latour, Bruno. *Reassembling the Social: An Introduction to Actor-Network-Theory.* Oxford, UK: Oxford University Press, 2005.

Lefebvre, Henri. *The Production of Space.* Translated by Donald Nicholson-Smith. Cambridge, MA: Blackwell, 1991.

Lin, Zhongjie. *Kenzo Tange and the Metabolist Movement: Urban Utopias of Modern Japan.* New York: Routledge, 2010.

Lockyer, Angus. "Japan at the Exhibition, 1867–1877: From Representation to Practice." In *Japanese Civilization in the Modern World XVII: Collection and Representation,* edited by Tadao Umesao, Angus Lockyer, and Kenji Yoshida, 67–76. Osaka: National Museum of Ethnology, 2001.

Loewy, Raymond. *The Locomotive.* New York: Studio, 1937.

Lyon, Peter. "Is This Any Way to Ruin a Railroad?" *American Heritage* 19, no. 2 (February 1968). https://www.americanheritage.com/any-way-ruin-railroad.

Ma, Alexandra. "Japanese Rail Company Issued an Official Apology After a Train Left 25 Seconds Early." *Business Insider*, May 15, 2018, https://www.business insider.com.

Maema Takanori. *Dangan ressha: Maboroshi no Tokyo hatsu Pekin yuki chōtokkyū.* Tokyo: Jitsugyō no Nihonsha, 1994.

Mainichi Shinbunsha, ed. *Supīdo 100-nen.* Tokyo: Mainichi Shinbunsha, 1969.

Marotti, William A. *Money, Trains, and Guillotines: Art and Revolution in 1960s Japan.* Asia-Pacific. Durham, NC: Duke University Press, 2013.

Marx, Leo. "The Impact of the Railroad on the American Imagination, as a Possible Comparison for the Space Impact." In *The Railroad and the Space Program: An Exploration in Historical Analogy,* edited by Bruce Mazlish, 202–216. Cambridge, MA: MIT Press, 1965.

Masumura Yasuzō, dir. *Kuro no chōtokkyū.* Daiei, 1964.

———. *Kuro no tesuto kā.* Daiei, 1962.

Matsusaka, Yoshihisa Tak. *The Making of Japanese Manchuria, 1904–1932.* Cambridge, MA: Harvard University Asia Center, 2001.

Mattelart, Armand. *The Invention of Communication.* Translated by Susan Emanuel. Minneapolis: University of Minnesota Press, 1996.

McDonald, Kate. "Asymmetrical Integration: Lessons from a Railway Empire." *Technology and Culture* 56, no. 1 (January 2015): 115–149.

———. *Placing Empire: Travel and the Social Imagination in Imperial Japan.* Oakland: University of California Press, 2017.

McKevitt, Andrew C. *Consuming Japan: Popular Culture and the Globalizing of 1980s America.* Chapel Hill: University of North Carolina Press, 2017.

Mettler, Meghan Warner. "Gimcracks, Dollar Blouses, and Transistors: American Reactions to Imported Japanese Products, 1945–1964." *Pacific Historical Review* 79, no. 2 (2010): 202–230.

Miller, Ian J. *The Nature of the Beasts: Empire and Exhibition at the Tokyo Imperial Zoo.* Berkeley: University of California Press, 2013.

Mimura, Janis. *Planning for Empire: Reform Bureaucrats and the Japanese Wartime State.* Ithaca, NY: Cornell University Press, 2011.

Minami Manshū Tetsudō Kabushiki Gaisha. *Ryūsenkei Tokkyū Ajia.* Dalian, China: Minami Manshū Tetsudō, 1936. Reprinted 1981.

Miyagi, Taizo. "Post-War Japan and Asianism." *Asia-Pacific Review* 13, no. 2 (2006): 1–16.

Mizuno, Hiromi. "Introduction: A Kula Ring for the Flying Geese: Japan's Technology Aid and Postwar Asia." In *Engineering Asia: Technology, Colonial Development and the Cold War Order,* edited by Hiromi Mizuno, Aaron S. Moore, and John DiMoia, 1–40. London: Bloomsbury, 2018.

Mooney, Suzanne. "Constructed Underground: Exploring Non-Visible Space Beneath the Elevated Tracks of the Yamanote Line." *Japan Forum* 30, no. 2 (2018): 205–223.

Moore, Aaron S. *Constructing East Asia: Technology, Ideology, and Empire in Japan's Wartime Era, 1937–1945.* Stanford, CA: Stanford University Press, 2013.

———. "From 'Constructing' to 'Developing' Asia: Japanese Engineers and the Formation of the Post-Colonial, Cold War Discourse of Development in Asia." In *Engineering Asia: Technology, Colonial Development and the Cold War Order,* edited by Hiromi Mizuno, Aaron S. Moore, and John DiMoia, 85–112. London: Bloomsbury, 2018.

Morley, David, and Kevin Robbins. *Spaces of Identity: Global Media, Electronic Landscapes and Cultural Boundaries.* New York: Routledge, 1995.

Morris-Suzuki, Tessa. *Beyond Computopia: Information, Automation, and Democracy in Japan.* New York: Kegan Paul International, 1988.

Morton, Ralph. "Grand Tour in Manchukuo." *The Spectator,* no. 5628 (May 8, 1936): 831–832.

Morton, T. Ralph, and Jenny Morton. Papers of T. Ralph Morton, 1924–1972. Centre for the Study of World Christianity, University of Edinburgh. GB 3189 CSCNWW11 on Archives Hub. https://archiveshub.jisc.ac.uk/data/gb3189-csc nww11.

Mrázek, Rudolf. *Engineers of Happy Land: Technology and Nationalism in a Colony.* Princeton, NJ: Princeton University Press, 2002.

Mumford, Lewis. "Authoritarian and Democratic Technics." *Technology and Culture* 5 (1964): 1–8.

Nagasawa Kikuya. *Shinkansen ryokō memo.* Tokyo: Shinju shoin, 1964.

Nakabō Kōhei. *Nakabō Kōhei: Watashi no jikenbo.* Tokyo: Shūeisha, 2000.

New Jersey State Highway Department, Division of Railroad Transportation. "The Japanese National Railway System and Its New Tokaido Line: A Report." Trenton, NJ: Bureau of Public Information, New Jersey State Highway Department, 1963.

"News and Comment." *Japan Architect* 36, no. 4 (1961): 7.

Neimark, Anna. "The Infrastructural Monument: Stalin's Water Works Under Construction and in Representation." *Future Anterior* 9, no. 2 (Winter 2012): 1–14.

Nihon Bōeki Shinkōkai, ed. *Nyū Yōku sekai hakurankai sanka hōkokusho 1964–1965.* Tokyo: Nihon Bōeki Shinkōkai, 1966.

Nishiguchi Katsumi. *Q-to monogatari.* Nishiguchi Katsumi shōsetsushū. Vol. 12. Tokyo: Shin Nihon Shuppansha, 1988.

———. *Shinkansen.* Tokyo: Kōbundō, 1966.

———. *Shinkansen.* Nishiguchi Katsumi shōsetsushū. Vol. 6. Tokyo: Shin Nihon Shuppansha, 1988.

Nishiyama, Takashi. *Engineering War and Peace in Modern Japan, 1868–1964.* Baltimore: Johns Hopkins University Press, 2014.

Noguchi, Paul H. *Delayed Departures, Overdue Arrivals: Industrial Familialism and the Japanese National Railways.* Honolulu: University of Hawai'i Press, 1990.

Nye, Joseph. *Soft Power: The Means to Success in World Politics*. New York: Public Affairs, 2006.

Oba Mie. "Japan's Entry into ECAFE." In *Japanese Diplomacy in the 1950s: From Isolation to Integration*, edited by Makoto Iokibe, et al., 98–113. New York: Routledge, 2008.

Oda, Meredith. "Masculinizing Japan and Reorienting San Francisco: The Osaka-San Francisco Sister-City Affiliation During the Early Cold War." *Diplomatic History* 41, no. 3 (2017): 460–488.

Orr, James Joseph. *The Victim as Hero: Ideologies of Peace and National Identity in Postwar Japan*. Honolulu: University of Hawai'i Press, 2001.

Osaka Jin. Tokyo: Asahi Shinbunsha, 1964.

Perez, Rafael Ivan Pazos. "The Historical Development of the Tokyo Skyline: Timeline and Morphology." *Journal of Asian Architecture and Building Engineering* 13, no. 3 (September 2014): 609–615.

Peters, John Durham. *The Marvelous Clouds: Toward a Philosophy of Elemental Media*. Chicago: University of Chicago Press, 2015.

Peterson, William. "Shinto and Cherry Blossoms: The Remasculinisation of Japan at the 1964–65 New York World's Fair." In *Exporting Japanese Aesthetics: Evolution from Tradition to Cool Japan*, edited by Tets Kimura and Jennifer Anne Harris, 80–104. Eastbourne, UK: Sussex Academic Press, 2020.

Prichard, Franz. *Residual Futures: The Urban Ecologies of Literary and Visual Media of 1960s and 1970s Japan*. New York: Columbia University Press, 2019.

Rance, Joseph, and Arei Kato. *Bullet Train*. New York: Morrow, 1980.

Rodgers, Dennis, and Bruce O'Neill. "Infrastructural Violence: Introduction to the Special Issue." *Ethnography* 13, no. 4 (2012): 401–412.

Roh, David S., Betsy Huang, and Greta A. Niu, *Techno-Orientalism: Imagining Asia in Speculative Fiction, History, and Media*. New Brunswick, NJ: Rutgers University Press, 2015.

Ross, Kerry. *Photography for Everyone: The Cultural Lives of Cameras and Consumers in Early Twentieth-Century Japan*. Stanford, CA: Stanford University Press, 2015.

Rydell, Robert W., John E. Findling, and Kimberly Pelle, *Fair America: World's Fairs in the United States*. Washington, DC: Smithsonian Books, 2000.

Said, Edward. *Orientalism*. New York: Pantheon Books, 1978.

Samuels, Richard J. *"Rich Nation, Strong Army": National Security and the Technological Transformation of Japan*. Ithaca, NY: Cornell University Press, 1994.

Sanuki Toshio. "Toshika jidai no kōtsū taikei." *Un'yu to keizai* 29, no. 2 (February 1969): 47–55.

Sasaki-Uemura, Wesley. *Organizing the Spontaneous: Citizen Protest in Postwar Japan*. Honolulu: University of Hawai'i Press, 2001.

Satō Jun'ya, dir. *Shinkansen daibakuha*. Tōei, 1975.

Satō Tatsurō. "Akikaze tatsu: Kane mōke shugi ni hihan no koe." *Sekai shūhō* 45, no. 40 (October 6, 1964): 60–63.

Schaller, Michael. *Altered States: The United States and Japan Since the Occupation.* New York: Oxford University Press, 1997.

Schivelbusch, Wolfgang. *The Railway Journey: The Industrialization of Time and Space in the 19th Century.* Berkeley: University of California Press, 1986.

Scott, James. *Weapons of the Weak: Everyday Forms of Peasant Resistance.* New Haven, CT: Yale University Press, 1987.

Seamon, David. "Body-Subject, Time-Space Routines, and Place-Ballets." In *The Human Experience of Space and Place,* edited by Anne Buttimer and David Seamon, 148–165. New York: St. Martin's Press, 1980.

Seraphim, Franziska. *War Memory and Social Politics in Japan, 1945–2005.* Cambridge, MA: Harvard University Asia Center, 2006.

Sert, J. L., F. Léger, and S. Giedion. "Nine Points on Monumentality." In *Architecture You and Me: The Diary of a Development,* edited by S. Giedion, 48–51. Cambridge, MA: Harvard University Press, 1958.

Shibusawa, Naoko. *America's Geisha Ally: Reimagining the Japanese Enemy.* Cambridge, MA: Harvard University Press, 2006.

Shimizu, Sayuri. *Creating People of Plenty: The United States and Japan's Economic Alternatives, 1950–1960.* Kent, OH: Kent State University Press, 2001.

Shimoda, Hiraku. "'The Super-Express of Our Dreams' and Other Mythologies About Postwar Japan." In *Trains, Culture, and Mobility: Riding the Rails,* edited by Benjamin Fraser and Steven D. Spalding, 263–289. Lanham, MD: Lexington Books, 2012.

Shinkansen '62. [Tokyo]: Nihon Kokuyū Tetsudō, 1962.

"Shinkansen are kore." *Dōboku gakkai shi* 49, no. 10 (1964): 42–53.

"Shinkansen ato ichi-nen: Zensen 515km o sora kara haiken suru." *Bungei shunjū* 41, no. 11 (1963): [n.p.].

Shinkansen o ichi nichi mo hayaku. [Tokyo]: Nihon Kokuyū Tetsudō, [1963].

"Shinkansen to Tokyo: Tokushū." Special issue, *Tokyo jin,* no. 282 (May 2010).

Shōgaku kokugo yomihon, vol. 10; reprinted in Kaigo Tokiomi, ed., *Nihon kyōkasho taikei kindai hen,* vol. 8, 150–153. Tokyo: Kōdansha, 1964.

Simone, AbdouMaliq. "People as Infrastructure: Intersecting Fragments in Johannesburg." *Public Culture* 16, no. 3 (2004): 407–429.

Smith, Kerry. "The Shōwa Hall: Memorializing Japan's War at Home." *Public Historian* 24, no. 4 (Autumn 2002): 35–64.

Sogō Shinji. "Gijutsu kakumei to Shinkansen." In *Sogō Shinji: Bessatsu,* 146–150. Tokyo: Sogō Shinji Den Kankōkai, 1988.

Solt, George. *The Untold History of Ramen: How Political Crisis in Japan Spawned a Global Food Craze.* Berkeley: University of California Press, 2014.

Spaeth, Ryu. "How *The New Yorker* Fell into the 'Weird Japan' Trap." *New Republic,* December 17, 2020. https://newrepublic.com/article/160595/new-yorker-japan -rent-family-fabricated.

Stoler, Ann Laura. *Along the Archival Grain: Epistemic Anxieties and Colonial Common Sense.* Princeton, NJ: Princeton University Press, 2008.

Suda Hiroshi. *Tōkaidō Shinkansen sanjū-nen.* Tokyo: Taishō Shuppan, 1994.

Suleski, Ronald. "Salvaging Memories: Former Japanese Colonists in Manchuria and the Shimoina Project, 2001–12." In *Empire and Environment in the Making of Manchuria,* edited by Norman Smith, 197–220. Vancouver: UBC Press, 2017.

Swenson-Wright, John. *Unequal Allies?: United States Security and Alliance Policy Toward Japan, 1945–1960.* Stanford, CA: Stanford University Press, 2005.

Tachibana Jirō. "Tokyo Shimonoseki aida Shinkansen ni tsuite." *Kōgyō zasshi* 76, no. 952 (April 1940): 166–169.

Tadokoro Masayuki. "The Model of an Economic Power: Japanese Diplomacy in the 1960s." In *The Diplomatic History of Postwar Japan,* edited by Iokibe Makoto, translated by Robert D. Eldridge, 81–107. New York: Routledge, 2011.

Takasaki Tatsunosuke. "Ajia no han'ei to Nihon no unmei." *Chūō kōron* 73, no. 1 (January 1958): 105–109.

Takenaka, Akiko. "Architecture as Wartime Cultural Diplomacy: The Japanese Pavilion at Paris 1937." In *Architecture of Great Expositions 1937–1959: Messages of Peace, Images of War,* edited by Rika Devos, Alexander Ortenberg, and Vladimir Paperny, 71–80. Burlington, VT: Ashgate, 2015.

Tamanoi, Mariko Asano. *Memory Maps: The State and Manchuria in Postwar Japan.* Honolulu: University of Hawai'i Press, 2009.

Tanaka, Kakuei. *Building a New Japan: A Plan for Remodeling the Japanese Archipelago.* Tokyo: Simul Press, 1973. Originally published as Tanaka Kakuei, *Nihon rettō kaizō ron* (Tokyo: Nikkan Kogyo Shinbunsha, 1972).

Tange, Kenzo. "Recollections: Architect Kenzo Tange (5)." *Japan Architect* 60, no. 8 (August 1985): 6–13.

———. "Recollections: Architect Kenzo Tange (6)." *Japan Architect* 60, no. 9 (September 1985): 6–15.

Tange, Kenzo, Koji Kamiya, Arata Isozaki, Sadao Watanabe, Noriaki Kurokawa, and Heiki Koh. "A Plan for Tokyo, 1960: Toward a Structural Reorganization." *Japan Architect* 36, no. 4 (April 1961): 8–34.

Tange, Kenzo, and Udo Kultermann. *Kenzo Tange, 1946–1969: Architecture and Urban Design.* New York: Praeger, 1970.

Tange Kenzō. "Dai Tōa kensetsu kinen eizō keikaku: Kyōgi sekkei tōsen zuan." *Kenchiku zasshi* 56, no. 693 (1942): 963–965.

———. "Nihon rettō no shōraizō: Tōkaidō megaroporisu no keisei." *Chiiki kaihatsu,* no. 2 (November 1964): 2–9.

Tetsudōshō Kansen Chōsaka. "Tokyo-Shimonoseki Shinkansen no zōsetsu." *Shūhō,* no. 192 (June 19, 1940): 35–39.

Time-Life Books. *Official Guide New York World's Fair 1964/1965.* New York: Time, 1964.

Togo, Kazuhiko. *Japan's Foreign Policy, 1945–2003: The Quest for a Proactive Policy.* Boston: Brill, 2005.

Tōkaidō Shinkansen. [Tokyo]: Nihon Kokuyū Tetsudō, [1964].

"Tokkyū 'Ajia' kaisō: Manshū o bakushin shita dangan ressha." In *Shōgen: Watashi no Shōwa-shi*, edited by Tokyo 12 Chaneru Hōdōbu, 124–135. Tokyo: Bungei Shorin, 1969.

Tollardo, Elisabetta. *Fascist Italy and the League of Nations, 1922–1935*. London: Palgrave Macmillan, 2016.

Tseng, Alice Y. *Modern Kyoto: Building for Ceremony and Commemoration, 1868–1940*. Honolulu: University of Hawai'i Press, 2018.

Ueno, Toshiya. "Japanimation and Techno-Orientalism." In *The Uncanny: Experiments in Cyborg Culture*, edited by Bruce Grenville, 223–236. Vancouver: Arsenal Pulp Press, 2002.

Umesao Tadao. "Bunmei kaeru chūsū shinkei: Shinkansen no 10-nen." In *Umesao Tadao chosakushū: Nihon kenkyū*, 402–403. Tokyo: Chūō kōronsha, 1990.

———. *Jōhō no bunmeigaku*. Tokyo: Chūō kōronsha, 1988.

"United States Delegations to International Conferences: ECAFE Study of Tokaido Railway." *Department of State Bulletin* 48, no. 1228 (January 7, 1963): 660.

Vaporis, Constantine Nomikos. *Breaking Barriers: Travel and the State in Early Modern Japan*. Cambridge, MA: Harvard University Press, 1995.

Volk, Alicia. "From Soft Power to Hard Sell: Japan at American Expositions, 1915–1965." In *JapanAmerica: Points of Contact, 1876–1970*, edited by Nancy E. Green and Christopher Reed, 66–87. Ithaca, NY: Herbert F. Johnson Museum of Art, Cornell University, 2016.

von Schnitzler, Antina. *Democracy's Infrastructure: Techno-Politics and Protest After Apartheid*. Princeton, NJ: Princeton University Press, 2016.

Walrath, Laurence K., and Curtis D. Buford. "Report of the United States Delegation to the United Nations Economic Commission for Asia and the Far East: Study Week of the New Tokaido Line, Tokyo, Japan (April 11 to April 18, 1963 Inclusive)." [1963].

Westad, Odd Arne. "The New International History of the Cold War: Three (Possible) Paradigms." *Diplomatic History* 24, no. 4 (2000): 551–565.

Wiener, Norbert. *Cybernetics; or, Control and Communication in the Animal and the Machine*, 2nd ed. Cambridge, MA: MIT Press, 1961.

———. *The Human Use of Human Beings: Cybernetics and Society*, 2nd ed. Garden City, NY: Doubleday, 1954.

Winner, Langdon. "Do Artifacts Have Politics?" *Daedalus* 109, no. 1 (Winter 1980): 121–136.

World Bank. *International Bank for Reconstruction and Development (World Bank) annual report 1961–1962*. Washington, DC: World Bank, 1962. http://documents .worldbank.org/curated/en/898261468764426372/International-Bank-for-Reconstruction-and-Development-World-Bank-annual-report-1961-1962.

Yamamoto Sae. *Senjika no banpaku to "Nihon" no hyōshō*. Tokyo: Shinwasha, 2012.

Yamamoto Sae, Aoki Fujio, Jessica Jordan, and Paul W. Ricketts. "From *The Representation of 'Japan* in Wartime World's Fairs: Modernists and 'Japaneseness.'" *Review of Japanese Culture and Society* 26 (December 2014): 104–134.

Yamanouchi Shūichirō. *Shinkansen ga nakattara*. Tokyo: Asahi Shinbunsha, 2004.

Yang, Daqing. *Technology of Empire: Telecommunications and Japanese Expansion in Asia, 1883–1945*. Cambridge, MA: Harvard University Asia Center, 2010.

Yatsuka, Hajime. "The 1960 Tokyo Bay Project of Kenzo Tange." In *Cities in Transition*, edited by Arie Graafland, 178–191. Rotterdam: 010 Publishers, 2001.

Yoneyama, Lisa. *Cold War Ruins: Transpacific Critique of American Justice and Japanese War Crimes*. Durham, NC: Duke University Press, 2016.

Yoneyama Toshinao. "Bunka kankō toshi." In *Kyoto no rekishi*, vol. 9, edited by Hayashi Tatsusaburō. Tokyo: Gakugei shorin, 1976.

Yoshimi, Shunya. "Information." Translated by David Buist. *Theory, Culture and Society* 23, no. 2–3 (2006): 271–278.

Young, Louise. *Japan's Total Empire: Manchuria and the Culture of Wartime Imperialism*. Berkeley: University of California Press, 1998.

Y-sei. "Mantetsu no hokori 'Ajia.'" *Kagaku Asahi* 2, no. 10 (1942): 28–29.

Yuasa Noriaki, dir. *Daikaijū Gamera*. Daiei, 1965.

Zwigenberg, Ran. *Hiroshima: The Origins of Global Memory Culture*. New York: Cambridge University Press, 2014.

Index

Adas, Michael, 2
Aguiar, Marian, 15
Ahuja, Ravi, 247n46
Amano Kōzō, 74
anti-communism, 186, 208, 213
Arima Hiroshi, 167
Asia Express: foreign views of, 149–53;
    name, 159–60; public memory of, 9,
    17–18, 139–44, 154–61, 170, 218, 220–
    21, 225–27; in railway imperialism,
    144–54; SMR employees, 153–54, 155,
    156–60; and wartime bullet train,
    164–65
Asia-Pacific War: public memory of
    high-speed rail and, 139–44, 159–78,
    221; and postwar foreign relations,
    180–84, 188, 206; World's Fairs dur-
    ing, 197
automated traffic control, 14, 102, 103,
    108, 110, 127; COMTRAC, 110, 135–
    37; foreign perceptions of, 18, 188,
    189, 194
automobiles, 3, 75, 92, 102, 134, 206
awards, 207
Avenell, Simon, 53

Benjamin, Walter, 96
Biwa, Lake, 26, 89, 92
Black Super-Express (*Kuro no chō-
    tokkyū*, film, Masumura Yasuzō),
    127–33
bridges, 12, 13, 189, 194, 264n116; Kamo
    River, 35; Lake Biwa Ōhashi, 92–
    93; Nagara River, 81; Seta River, 89;
    Yodo River, 78–79
Bullet Train (*Dangan ressha*, NHK tele-
    vision series), 82–87, 174–78, 224
"bullet train" (*dangan ressha*), wartime
    plan: and empire, war, 161, 162–66,
    169–70; as precursor to Tōkaidō
    Shinkansen, 170–71; public memory
    of, 139–44, 161–62, 169–79; tunnels,
    164, 166–69, 172–78

Carse, Ashley, 183
central traffic control. *See* automated
    traffic control
centralization, 64, 67, 72–76, 80, 101,
    104, 110–26. *See also* cultural
    homogenization
Chikaraishi Sadakazu, 74

36–39, 58, 63, 224; Technical Tourism Program, 191–93; through a train window, 89–90, 155; transportation infrastructure as object of, 11–12, 79, 89, 92–93, 191–93, 226–27. *See also* Japan Travel Bureau
trade policies, U.S., 180, 207–14
Trumbull, Robert (*New York Times* articles), 192, 198, 201, 202, 203
Tseng, Alice, 29–30
Tsuji Shūji, 37
tunnels: in fiction, 86–87, 167–68, 189, 194, 224; maglev, 224; New Tanna, 11, 14, 167–70, 172–78; significance, 12, 161, 165–68, 264n116; undersea, 18, 164, 168

Umesao Tadao, 106–9, 116, 137, 244n5–6
United Nations Economic Commission for Asia and the Far East. *See* ECAFE
urban planning. *See* development planning
U.S.-Japan Security Treaty (Anpo) protests: and democracy, 6, 40, 105, 111, 118, 141, 182, 234n6; model for local protests 22, 60, 221, 238n100
urban planning. *See* development planning
urbanization. *See* centralization

Wiener, Norbert, 110–11, 113, 115, 117, 245n26
Winner, Langdon, 6, 104, 244n3
World Bank, 36, 185, 257n17
World War II. *See* Asia-Pacific War
World's Fairs, 9, 180, 195–208, 211–12, 227–28, 259n58

Yamanote Line, 68, 71
Yatsuka, Hajime, 118
Yodo River, 78–79
Yokohama, 10, 67–68, 76–78, 95, 131–34, 192
Yoshimi, Shunya, 126
Young, Louise, 145
Yūrakuchō, 71

Zwigenberg, Ran, 234n11

STUDIES OF THE WEATHERHEAD
EAST ASIAN INSTITUTE
COLUMBIA UNIVERSITY

Selected Titles
(Complete list at: weai.columbia.edu/content/publications)

Middlemen of Modernity: Local Elites and Agricultural Development in Modern Japan, by Christopher Craig. University of Hawaiʻi Press, 2021.

Isolating the Enemy: Diplomatic Strategy in China and the United States, 1953–1956, by Tao Wang. Columbia University Press, 2021.

A Medicated Empire: The Pharmaceutical Industry and Modern Japan, by Timothy M. Yang. Cornell University Press, 2021.

Dwelling in the World: Family, House, and Home in Tianjin, China, 1860–1960, by Elizabeth LaCouture. Columbia University Press, 2021.

Disunion: Anticommunist Nationalism and the Making of the Republic of Vietnam, by Nu-Anh Tran. University of Hawaiʻi Press, 2021.

Made in Hong Kong: Transpacific Networks and a New History of Globalization, by Peter Hamilton. Columbia University Press, 2021.

China's Influence and the Center-periphery Tug of War in Hong Kong, Taiwan and Indo-Pacific, by Brian C.H. Fong, Wu Jieh-min, and Andrew J. Nathan. Routledge, 2020.

The Power of the Brush: Epistolary Practices in Chosŏn Korea, by Hwisang Cho. University of Washington Press, 2020.

On Our Own Strength: The Self-Reliant Literary Group and Cosmopolitan Nationalism in Late Colonial Vietnam, by Martina Thucnhi Nguyen. University of Hawaiʻi Press, 2020.

A Third Way: The Origins of China's Current Economic Development Strategy, by Lawrence Chris Reardon. Harvard University Asia Center, 2020.

Disruptions of Daily Life: Japanese Literary Modernism in the World, by Arthur M. Mitchell. Cornell University Press, 2020.

Recovering Histories: Life and Labor after Heroin in Reform-Era China, by Nicholas Bartlett. University of California Press, 2020.

Figures of the World: The Naturalist Novel and Transnational Form, by Christopher Laing Hill. Northwestern University Press, 2020.

Arbiters of Patriotism: Right Wing Scholars in Imperial Japan, by John Person. University of Hawai'i Press, 2020.

The Chinese Revolution on the Tibetan Frontier, by Benno Weiner. Cornell University Press, 2020.

Making It Count: Statistics and Statecraft in the Early People's Republic of China, by Arunabh Ghosh. Princeton University Press, 2020.

Tea War: A History of Capitalism in China and India, by Andrew B. Liu. Yale University Press, 2020.

Revolution Goes East: Imperial Japan and Soviet Communism, by Tatiana Linkhoeva. Cornell University Press, 2020.

Vernacular Industrialism in China: Local Innovation and Translated Technologies in the Making of a Cosmetics Empire, 1900–1940, by Eugenia Lean. Columbia University Press, 2020.

Fighting for Virtue: Justice and Politics in Thailand, by Duncan McCargo. Cornell University Press, 2020.

Beyond the Steppe Frontier: A History of the Sino-Russian Border, by Sören Urbansky. Princeton University Press, 2020.

Made in United States
North Haven, CT
29 March 2023

34719280R10183